Mariana de Oliveira Pereira
Carlos Alberto Vieira de Azevedo
Jailton Garcia Ramos

Growing fodder palm in the semi-arid region under saline conditions

Mariana de Oliveira Pereira
Carlos Alberto Vieira de Azevedo
Jailton Garcia Ramos

Growing fodder palm in the semi-arid region under saline conditions

Agronomic performance of forage palm under irrigation rates and levels of irrigation water salinity

ScienciaScripts

Imprint

Cover image: www.ingimage.com

This book is a translation from the original published under ISBN 978-620-4-19558-2.

Publisher:
Sciencia Scripts
is a trademark of
Dodo Books Indian Ocean Ltd. and OmniScriptum S.R.L publishing group

120 High Road, East Finchley, London, N2 9ED, United Kingdom
Str. Armeneasca 28/1, office 1, Chisinau MD-2012, Republic of Moldova, Europe
Printed at: see last page
ISBN: 978-620-8-07753-2

AGRONOMIC PERFORMANCE OF FORAGE PALM UNDER IRRIGATION RATES AND WATER SALINITY LEVELS

Authors:

MARIANA DE OLIVEIRA PEREIRA

CARLOS ALBERTO VIEIRA DE AZEVEDO

JAILTON GARCIA RAMOS

DEDICATORY

I dedicate the realisation of this dream and this achievement to my parents Marisis de Oliveira Pereira and Adilson Santos Pereira for all their love, encouragement, understanding and for always believing in my potential.

I would also like to dedicate this to my grandmothers Hermínia Cunha de Oliveira *(in memoriam)*, for always teaching me the importance of studying, and Nair Santos Pereira *(in memoriam)* for the joy of life that accompanies me in everything I do.

ACKNOWLEDGEMENTS

First and foremost, I thank God for all his providence, love and mercy towards me at every moment of my life. I thank him for giving me the wisdom, strength and discernment to make my choices and support me in them, so that I could reach this moment of graduating from the Postgraduate Programme in Agricultural Engineering: finally a doctorate. Thank you, Merciful Jesus, for this dream come true. It was never easy to leave everything behind and venture into an unknown land, but the Lord sustained me at every moment and placed special people in my life throughout this journey so that I could never feel alone. I would like to thank my mother, Our Lady, in all her denominations, and especially Our Lady of Aparecida and Our Lady of Grace, for her intercession and for placing me on her lap so many times.

I would like to thank my parents Adilson Santos Pereira and Marisis de Oliveira Pereira for all their affection, love, patience, teachings, unity, encouragement and so many other things that only you, with your infinite love for me, are capable of doing. Thank you for never letting me give up on anything I've done. You are my greatest examples and strength, after God. I know how difficult the distance is and I apologise for not always being able to be there, giving you the support I would like. This achievement is ours, this title is yours too. Thank you for everything. I love you.

I would especially like to thank my brother Victor Hugo de Oliveira Pereira for giving me the best gifts of my life: Nicholas José Couto Pereira and Giovana Maria Couto Pereira, my auntie's loves. Whenever I have difficult days, I think of my little ones and everything becomes much better.

I would like to thank all my family - Oliveira and Pereira - for their encouragement, affection, prayers and joy. Unfortunately, it's not possible to mention each and every one of you because, in that case, it would be the equivalent of writing a new thesis, as big as the family is. So I'd like to thank the Oliveira family, especially Aunt Márcia, Aunt Marcy, Uncle Francisco and Cousin Leila for their support and prayers. And on the Pereira family side, especially aunt Ana, godfather Adauto, cousins Sheylla, Jaqueline, Vaninha and Bárbara (who even helped me type the thesis data together with Dudu) for all the words of encouragement. Family, please don't be jealous if I've left someone out. Everyone is special in my life!

I would like to thank Professor Carlos Alberto Vieira de Azevedo (UFCG) for his guidance, patience and availability. Thank you for your understanding, teachings, support, conversations, corrections and demands. This has contributed so much to my professional and personal life.

To Professor and supervisor Gustavo Bastos Lyra (UFRRJ), thank you for accepting my guidance once again. You've been with me since my master's degree, you encouraged me to try my doctorate at UFCG and you made me passionate about agrometeorology (I don't know if that's good or bad, because working with agrometeorology and fodder palm wasn't easy at all). I told you during my master's degree that you wouldn't get rid of me that easily, remember?

Thank you for believing in my potential and my ability to go further. Coming to UFCG changed my life!

I would like to thank Professor Luciano Saboya Fallé above all for his friendship. Thank you for often treating me with the concern of a father. Countless times you devoted your time to helping me, to clearing up my doubts and to simply chatting away. I know you'll always remember me saying: oh, Professor, I'm not going to do that! My eternal gratitude to you.

I would like to thank Professor José Dantas Neto for his willingness to help me in any way he could, whether with material for the experiment or in an intellectual way.

To Professor Hugo Orlando Carvallo Guerra, thank you for helping me with my many doubts about the soil water balance and for your affection.

I would like to thank Professors Geovani Soares de Lima and Lauriane Almeida dos Anjos Soares for their friendship, availability, support and patience. I admire you!

I would like to thank Professor Ronaldo Nascimento for making the Physiology Laboratory available and for his contributions.

To Professor Vera Lúcia Antunes de Lima, thank you for your concern and support.

To all the professors of the Postgraduate Programme in Agricultural Engineering not mentioned here, the Federal University of Campina Grande, the Coordination for the Improvement of Higher Education Personnel and the National Semi-Arid Institute (INSA) for all the support needed to carry out this research.

To INSA researcher Jucilene Silva Araújo for her availability and support.

To the entire panel of judges for their contributions to the improvement of this work.

To Professor Carlos Pereira of the Fluminense Federal University and other colleagues from the same institution who always supported me.

I've always been asked how I managed to stay so far away from home in Rio de Janeiro. My answer is that God has placed people in my life who have accompanied me on this journey and whom I will always carry with me in my heart, wherever I go. Thank you to my "former interns" who are now great friends: João Henrique Cabral (and his family) and Beatriz Tomaz (and her family). I'm so glad UFCG gave you to me! Thank you to Kheila who was also by my side, helping whenever possible.

Thank you for the welcome and all the help you've given me since the beginning, without which it would have been impossible to continue: Rafa, Dani, Sah, Kecinha, Thiago, Tatá, Bruno thank you for looking after me. Benevenuto, Jean (both of them), Robson, Paty and Rigo and the boys from the salinity team (André, Adaan, Ellyson, Saulo, Carantino, Wesley and all the boys who helped me and made me happy, please forgive me if I've forgotten anyone). I know that this was the research that got the most people involved, I always had an activity for someone to help me with. Everything you did for me was incredible. With all my heart and eternal gratitude.

I would like to thank Elka Costa for her friendship, affection and partnership both inside and outside UFCG. I would also like to thank Carlos Vailan for his partnership and for everything he has done for me. I will miss our trips, studies and outings. And please, always remember to fill up the car on your next trips, "TANK" (I know you'll understand)!

I need to open a paragraph for the two of them, partners in the cave (our affectionate nickname for our living room)... Márcia, thank you for your affection, your companionship, your friendship, your nudges, in short, thank you for everything. Please know that this feeling of gratitude extends to your family. Thank you for helping me at all times and for being present in my life. To Garcia (better known as Jailton), I don't even know how to begin to thank you. Thank you above all for your sincere friendship, for everything you've done and are doing for me, for all the support, love and concern, for welcoming me into your home with your family, for our laughs over leseiras and for so many other things that don't fit on Lattes. You were the first person I met when I arrived at UFCG, and I will always carry you with me, my great friend. Thank you both for everything. It's impossible to describe my gratitude in words.

Cris girl, you're part of my life. No matter how far apart we are or how long we don't speak to each other. Thank you for always being by my side at all times. My partner through thick and thin. We'll always be the café au lait duo!

Fernandinha, you don't exist. Your heart is so generous. What you have for a little one, you have for a brave one when it comes to pulling my ears. Thank you for a friendship that doesn't judge, but corrects and stands up for others. Thank you for everything you've done for me, only we know!

Gabi, thank you for your love and concern. Our Goiás will have to stay with you for the time being, but maybe one day God will bring us together?

I would like to thank all the UFCG staff who were always so helpful, especially Mr Antônio and the boys from the transport department: Alda, Roberto, Michelle, Mrs Cida, Marciano and Mr Edil.

To Ronaldo Filho and his family, thank you for welcoming me with such affection.

Thank you to the Encounter with Christ Church of Our Lady of the Rosary for putting special people in my life too (Raquel, Tatá, Jaque and Anna).

To the Dominican Sisters Cleide, Silvana and Ana, you are examples of kindness, love and dedication to others. Thank you for everything!

To the women farmers of the Quintais Productivos and Banco de sementes projects, you are an inspiration to me in terms of strength, wisdom, determination and generosity. I would like to thank Mrs Luzia *(in memoriam)* for often welcoming me like a daughter with her love.

Finally, I would like to thank everyone who participated directly or indirectly in this achievement. I apologise if I've forgotten anyone, but that doesn't mean they're any less important. May God enlighten the lives of each and every one of you.

LIST OF TABLES

LIST OF FIGURES

of water salinity (ECa) for each irrigation rate (L1 = 25, L2 = 50, L3 = 75, and L4 = 100% of total soil water capacity - CTA) at 442 days of the cycle.

Figure 16. Number of primary cladodes (NCP) of Mexican Elephant Ear forage palm as a function of irrigation rates at 442 days into the cycle.

Figure 17. Fresh mass of primary cladode (MFCP) of Mexican Elephant Ear forage palm as a function of irrigation rates and electrical conductivity of irrigation water (ECa) at 442 days of the cycle.

Figure 18. Fresh mass of secondary cladodia (MFCS) of Mexican Elephant Ear forage palm as a function of water salinity (ECa) for each irrigation regime (L1 = 25, L2 = 50, L3 = 75, and L4 = 100% of total soil water capacity - CTA) at 442 days of the cycle 60

Figure 19. Total fresh mass (FFM) of Mexican Elephant Ear forage palm as a function of irrigation rates at 442 days into the cycle.

Figure 20. Total fresh mass (MFT) of Mexican Elephant Ear forage palm as a function of irrigation water salinity levels (ECa) at 442 days of the cycle.

SUMMARY

Agronomic performance of fodder palm under irrigation rates and water salinity levels

ABSTRACT: Although forage palm is adapted to the Brazilian semi-arid region, it can be influenced by the edaphoclimatic conditions of this region. The aim of this research was to evaluate the effect of different irrigation rates and water salinity levels on the cultivation of the Mexican Elephant Ear species. The research was conducted in open-air pots at the Federal University of Campina Grande between September 2017 and December 2018. The experimental design was randomised blocks in a 4 x 4 factorial scheme, with 4 replications. The factors consisted of four irrigation rates (25, 50, 75 and 100 per cent) as a function of the total water capacity in the soil and four levels of electrical conductivity of the water: 0.60, 3.00, 5.40 and 7.80 dS m^{-1}

Reference evapotranspiration (ETo), actual evapotranspiration (ETr) and the crop coefficient (kc) were determined. Morphometric, physiological and biomass crop variables were also assessed. Applying irrigation rates in the 320.00 to 460.00 mm range favours the ETr of fodder palm and even under saline conditions the plant maintains its evapotranspirometric demand. The kc varied between 0.68 and 0.85. The growth of the crop was not affected by irrigation rates and salinity levels, except for the thickness of the secondary cladode. Increasing salinity reduced the plant's water content and increased the percentage of extravasated electrolytes. The Mexican Elephant Ear palm's threshold salinity of 2.23 dS m^{-1} and the 354.00 mm blade resulted in higher total fresh mass production values of 9.34 and 5.94 kg respectively.

Keywords: *Opuntia stricta* (Haworth) Haworth; water availability; electrical conductivity of water.

1. INTRODUCTION

The semi-arid region is characterised by scarce and irregular rainfall, high annual evaporation of more than 2000 mm and shallow soils with low moisture retention capacity, limiting agricultural activities in the Northeast of Brazil (SILVA et al., 2014b). These conditions lead to seasonality in crop production during the dry season, and this includes a reduction in the availability of fodder for animal feed. In this scenario, the Mexican Elephant Ear palm *(Opuntia stricta (*Haworth) Haworth) stands out as a forage plant capable of mitigating the effects of low livestock yields. This is because the efficiency with which the *Opuntia* genus uses water from the soil *is* around 100 to 150 litres of water for every kilogram of dry matter produced, while grasses need between 250 and 350 litres of water to produce the same amount of dry matter (LARCHER, 1986).

Forage palm species are cacti with the greatest potential for exploitation in the Northeast of Brazil and are an important forage resource during periods of drought, due to their high phytomass production potential in the environmental characteristics of the semi-arid region (CAMPOS et al., 2017). Despite being species adapted to this region, local meteorological conditions have a strong influence on plant development, as water deficit can promote a reduction in water content and water potential and result in loss of turgidity, closure of stomata, reduced growth and, consequently, a decrease in final production. In this way, the realisation of irrigation events becomes a preponderant practice for the production system and the evapotranspiration process must be taken into account, as well as the plants' response to the different conditions of water availability in the soil (SILVA et al., 2012).

Although water is a limiting factor for agricultural production in the semi-arid region, it is necessary to look for better alternatives for crop development in order to optimise the use of water, based on determining water requirements in accordance with the soil and climate characteristics of each region. Obtaining estimates of the crop's evapotranspiration allows the total amount of water required by the plant to be determined and provides adequate irrigation management, favouring increased production.

The use of lower quality water has been the alternative found by many producers in the Northeast to minimise the effects of water scarcity on plants. However, it should be borne in mind that the quality of the water available for irrigation in various locations in the Brazilian semi-arid region contains high levels of soluble salts. In these circumstances, the accumulation of salts in the soil can cause the plant's water needs to change due to alterations in the process of water absorption and evapotranspiration, as well as further contributing to soil degradation (TOLEDO et al., 2017). According to Ribeiro et al. (2016a), salinity is one of the abiotic stresses that most limits agricultural production due to its negative effects on plant growth and development.

The sensitivity of crops to salts in water and soil highlights the need for research aimed at obtaining viable technologies for producers that can minimise the effects of salinity on plants (SOUSA et al., 2012).

Knowledge of the water consumption of oil palm makes it possible to define the appropriate irrigation management in order to maximise the final production and thus contribute to the rural producer's income. Studying the effects of salinity on water consumption, growth, physiology and production of forage palm *(Opuntia stricta* (Haworth) Haworth) is of great importance, given that this forage has been used as an alternative for animal feed by farmers in the semi-arid region during times of drought. Understanding the effects of salts in water and soil on oil palm helps to assess the feasibility of using saline water in agriculture to grow it. The use of saline water also contributes to saving water in regions with low water availability by using lower quality water for irrigation, while maintaining good quality water for more noble purposes such as human consumption and animal watering.

With regard to fodder palm, some authors (SALES et al., 2013; SILVA et al., 2015a; LIMA et al., 2016; CRUZ NETO et al., 2017; AMORIM et al., 2017) have carried out studies to assess the crop's water relations, growth and productivity under different management conditions. However, there are few studies covering the crop's water needs throughout its cycle, depending on the application of irrigation rates and relating the effects of irrigation water salinity on the plant.

Thus, there is a need for studies aimed at improving fodder palm production systems and keeping farmers in the countryside, as well as the use of saline water in agriculture, given the current scenario of water scarcity, and the technical and scientific development of students and researchers.

2. OBJECTIVES

2.1 General objective

The general aim of this research was to evaluate the effect of applying irrigation rates and levels of salinity in irrigation water on water consumption, growth, physiology and production of Mexican Elephant Ear palm *(Opuntia stricta (*Haworth) Haworth).

2.2 Specific objectives

- Determine reference evapotranspiration (ET_o) using the Penman-Monteith - FAO 56 method;
- Estimating the actual evapotranspiration (ET_r) of the crop using the gravimetric soil water balance;
- Determine the single crop coefficient of fodder palm (k_c);
- Evaluate morphometric, physiological and biomass parameters of forage palm and;
- Determine the threshold salinity of fodder palm.

3. THEORETICAL FRAMEWORK

3.1 Relevance of fodder palm in north-eastern Brazil

Livestock farming in the semi-arid region of Brazil is influenced by the temporal and spatial variability of rainfall, which has led to a search for forage plants that are adapted to local climatic conditions. Among these plants, the one most commonly used for animal feed in north-eastern Brazil is forage palm, a xerophilous species that, due to its morphophysiological characteristics, has a certain tolerance to environments with water scarcity, high temperatures and low fertility soils (PEREIRA et al., 2015b).

Palm is a species that stands out in the Brazilian semi-arid region compared to other forage crops, as it is part of the dietary base for goats and sheep, which are of great social and economic importance to producers in the region. This crop is an important forage resource, helping to meet the demand for food during the dry season due to its hardiness, high production potential, high palatability and high energy value when compared to other species of native vegetation. Palm can be used as a total or partial substitute for other ingredients with higher production costs, such as maize. It also makes a valuable contribution to the animals' water supply as it contains an average of 90 per cent water in its cladodes. Research evaluating ruminants consuming diets with palm inclusion of more than 50 per cent found a reduction in water consumption of 87 per cent in sheep and 74 per cent in Holstein cows (FRANCISCO et al., 2015; LIMA et al., 2015).

It is estimated that the Brazilian semi-arid region has an area cultivated with this cactaceae of over 600,000 ha for fodder production (FROTA et al., 2015). Among the most cultivated species are *Opuntias*, with around 40,000 ha of this plant in the northeastern states of Paraíba, Pernambuco and Alagoas. The cost of healthy seed racks for planting varies between R$ 0.20 and 0.40 per unit. If planted at a density of 40,000 plants per hectare, fodder palm can surpass corn in terms of energy production per hectare. And in a production of 350 t of green matter ha^{-1} $year^{-1}$ the crop provides 315,000 litres of water. With the severe shortage of water for animals in the dry season, this characteristic of palms is of strategic importance (LIMA et al., 2015).

The reason for the increased interest in *Opuntias*, and in particular *Opuntia stricta* (Haworth) Haworth, is the important role they can play in the success of sustainable agricultural systems in arid and semi-arid zones. This is due to its high degree of resistance to drought and high temperatures, its adaptability to poorly fertile soils, its high productivity as a result of its high water use efficiency, as well as the economic role it can play in increasing the viability and economic efficiency of small and medium-sized plots of low-income farmers seeking to produce for their subsistence (FAO, 1995).

3.2 General aspects of the crop

Forage palm species belong to the family Cactaceae Juss. In this family, there are 178 genera with around 2,000 known species. The most widely cultivated species of fodder palm for animal feed include: Miúda or Doce *(Nopalea cochenillifera* Salm Dyck), Baiana or IPA-Sertânia *(Nopalea cochenillifera* Salm Dyck) and Orelha de Elefante Mexicana *(Opuntia stricta* (Haworth) Haworth). In general, cultivars of the *Opuntia* genus tolerate more intense droughts and are more resistant to the scale insect *(Diaspis echinocacti* Bouché) than the *Nopalea* genus. This genus, in turn, is more resistant to the carmine mealybug *(Dactylopius opuntiae* Cockerell) (VASCONCELOS et al., 2009).

Plants of the species *Opuntia stricta* (Haworth) Haworth belong to the division Embryophyta, subdivision Angiospermea, class Equisetopsida C. Agardh, subclass Magnoliidae Novák ex Takht., family Cactaceae and genus *Opuntia* (TROPICOS, 2020). It is a succulent and thorny species native to the southeastern United States, eastern Mexico and some Caribbean islands (SHACKLETON et al., 2017). They are of the arborescent type with a height of more than 1 m; broad crown; glabrous; obovate cladodes 30-60 cm long, 20-40 cm wide, dark green, covered with a layer of wax (ZAPPI et al., 2015). The cladode is an ellipse formed by: skin, bark, a ring of vascular tissue made up of collateral bundles separated by parenchymatous tissue, and a pith which is the main succulent tissue (TING, 1988; FAO, 1995). According to the same authors, the stomata are evenly distributed over both sides of the cladode surface and are randomly dispersed, not being very numerous (usually 15 to 35 stomata per mm^2). The flowers are yellow in colour and 25-30 mm long. The fruit is sweet, juicy and edible, 40-60 mm long and 25-30 mm wide (PINKAVA, 2004). It has a system of superficial, fleshy roots with a horizontal distribution. The distribution of the roots may depend on the type of soil and the management of the plantation. Under favourable soil conditions, an extended root develops, which penetrates almost 30 cm into the soil; under drought conditions, such as those that occur in arid and semi-arid regions, fleshy lateral roots develop from the main root with the function of absorbing water at low levels. However, in all types of soil, the mass of absorbing roots is found in the first few centimetres, with a maximum depth of 30 cm and a spread of 4 to 8 m (TING, 1988; FAO, 1995).

The morphophysiological aspects that make the palm an interesting option for arid and semi-arid zones are linked to its impermeable cuticle, smaller number of stomata and photosynthetic apparatus. Its *crassulacean acid* metabolism *(*CAM) has the ability to capture solar energy during the day and fix CO_2 at night, reducing water loss through transpiration. In this way, it is more efficient at using water, being 10 times more efficient than a plant with a C3 photosynthetic metabolism (ROCHA, 2012). Transpiration involves the evaporation of water inside the photosynthetic organs (cladodes), whose internal air spaces are saturated with water vapour. The pores of the stomata of *O. ficus-indica* occupy only 0.50 per cent of the plant's surface, even when they are completely open, while the stomata of the leaves of highly productive C3 or C4 plants occupy two to three times more leaf area. During a 24-hour period, *Opuntias* can transpire 11.3 moles (203 g) of water per m^2 of surface area, while a highly productive C4 plant can lose around 2.9 times as much and a C3 plant up to 4.7 times as much (FAO, 1995). However, Silva et al. (2015b) state that some CAM species adjust their CO_2 capture pattern to C3 plants when subjected to higher water availability conditions.

The productive potential of forage palm species occurs in regions where the average temperature ranges between 16.1 °C and 25.4 °C; with highs between 28.5 °C and 31.5 °C and lows ranging from 8.6 °C to 20.4 °C. The temperature range is between 10.0 °C and 17.2 °C. The ideal rainfall range is between 368.4 mm and 812.4 mm, although it can be grown with 200 mm, and relative humidity at night is between 55 and 60 per cent (ROCHA, 2012).

The first cut of forage palm managed under rainfed conditions takes place two years after planting and subsequent cuts are made every two years. In an irrigated cultivation system, the first cut can be made at 12 months and subsequent cuts according to forage requirements (LIMA et al., 2015).

Studies conducted with forage palm species show that although the crop tolerates extreme drought conditions, the application of supplementary irrigation events favours productive performance and thus contributes to increasing food availability, especially during the dry season (BARBOSA et al., 2017). According to Santos et al. (2006), oil palm can reach yields of up to 40 tonnes of dry matter per hectare in rainfed cultivation and the productive benefits of applying water to the plants can be seen more clearly in regions with low rainfall levels. However, scientific information on growing fodder palm under irrigation systems is still incipient.

According to Oliveira et al. (2010), there is little knowledge about the water requirements of oil palm in order for it to achieve high production. In the literature, most of the studies on forage palm are on the *Opuntia ficus-indica* species, *and* there are few studies on vegetative production with the *Opuntia stricta* (Haworth) Haworth species.

3.3 Biosaline agriculture

Increasing the productive potential of the *Opuntia stricta* (Haworth) Haworth forage palm associated with the use of irrigation systems, although contradictory for water-scarce regions, should be considered. To this end, the use of saline water in agriculture should be considered in the light of the quality of the water sources available in arid and semi-arid areas, which generally have a high concentration of solutes as a result of the processes of eutrophication and salinisation. The salinisation process involves the concentration of soluble salts in the soil solution and results in the formation of saline soils with the accumulation of Na^+, Ca^2 +, Mg^2 + and K+ in the soil horizons. The main salts found are chlorides and sulphates of Na^+, Ca^2 +, Mg^2 and carbonates and nitrates are found in small quantities (RESENDE et al., 2014; PEDROTTI et al., 2015).

In the north-east of Brazil, most areas are salinised due to limited rainfall, high evapotranspiration demand, poor drainage and the use of irrigation water with a high salt content. According to Gheyi et al. (2010), the Brazilian semi-arid region has an average annual rainfall of 750 mm, although in some areas the average rainfall does not exceed 400 mm per year. Average potential evapotranspiration reaches 2,500 mm per year, generating high water deficits. This deficit favours the concentration of solutes in water sources, degrading water quality. The predominance of crystalline rocks in the subsoil in much of the semi-arid region imposes brackish and saline characteristics on groundwater, making it difficult to use for human consumption. Despite these limitations, these waters are important from a social point of view, as they can meet the needs of families who use them for irrigation and animal husbandry.

In the state of Paraíba, in the north-east of Brazil, the marked climatic conditions due to both spatial and temporal irregularities in the rainfall regime (FRANCISCO et al., 2015), combined with other factors, lead to a reduction in forage productivity. Therefore, an alternative to minimise the deleterious effects of lack of rainfall on crops is the use of irrigation systems. The use of saline water should be considered in irrigation in view of the scarcity of water resources. However, the management of saline water in agriculture must be done in such a way that it does not salinise the soil or harm the crop. According to Lima et al. (2015), satisfactory results from the use of saline water in fodder palm irrigation are associated with good drainage, depth and sand content (> 75%) in the soils, associated with the small amount of water used and, consequently, salts, which are washed away annually with the arrival of the rains.

It is in this context that we see the importance of developing biosaline agriculture, which consists of using saline water, usually in saline or salinised soils, to grow species that are tolerant to saline stress (halophytic species), when good quality water and soils become scarce, through the application of appropriate irrigation methods (GHEYI et al., 2010; OLIVEIRA et al., 2017).

3.3.1. Crop tolerance to salinity

Although forage palm species are not considered to be tolerant to salinity, according to Lima et al. (2015), biosaline agriculture is already a reality for producers in this region, given that these waters have been used to irrigate this cactaceae in experimental fields and private establishments, with yields of over 20 tonnes of dry matter per hectare per year^{-1} of *Opuntia* and *Nopalea.* The authors also state that scientific information on the use of saline water and its consequences on the different types of soil in the semi-arid region also needs to be obtained for palm cultivation.

Soil salinisation has negative effects on plants due to the increase in the soil's osmotic potential, which results in greater energy expenditure by the plant in the process of absorbing water from the soil and other vital elements; due to ionic toxicity, mainly sodium, boron, bicarbonates and chlorides, which in high concentrations cause physiological disturbances in plants. Salinisation can also cause nutritional imbalance due to the significant alteration in the processes of absorption, transport, assimilation and distribution of nutrients in the plant. Changes in soil pH values also reduce the availability of micronutrients. Salinity negatively affects water absorption by the roots. The presence of salts in the soil solution leads to a decrease in the external water potential. The water deficit in plant tissues as a result of the osmotic effect can cause stomatal closure to reduce water loss through transpiration, consequently reducing the photosynthetic rate, which affects plant development and productivity throughout its cycle (GHEYI et al., 2010; SILVA et al., 2013a; PEDROTTI et al., 2015).

Therefore, the use of saline water to irrigate fodder palm requires further study. According to Freire (2012), the palm has no tolerance to saline stress. Under conditions of salinity, plant growth is reduced due to changes in ion intake and metabolism, water potential, mineral nutrition, stomatal closure and photosynthetic efficiency. However, Lima et al. (2015) when using saline water to irrigate oil palm, with an electrical conductivity of more than 5 dS m^{-1} and considered to be of high salinity and high chloride levels, found that for a period of six years productivity was not affected and there was no damage to the soil. However, the literature has yet to define threshold salinity levels for the different forage palm genera.

3.4 Water consumption of fodder palm

Another fundamental factor for irrigating fodder palm, among other crops, is knowledge of water consumption during its cycle, both in terms of boosting productivity and rationalising the use of water resources. However, although irrigation is seen as beneficial to increasing palm production, excess water in the soil can jeopardise production, given that this is a crop adapted to semi-arid conditions. In this way, excessive irrigation can impair soil aeration and favour the development of plant diseases, especially rot, due to the action of fungi and bacteria. In the Elephant's Ear forage palm *(Opuntia stricta* (Haworth) Haworth), high relative humidity and rainfall favour the intensity of disease in plantations (LIMA et al., 2017; SOUZA et al., 2017). Therefore, information that characterises the response of plants to different levels of water availability is important for defining the best ways to treat the disease.

management strategies aimed at maximum water utilisation by the crop (PEREIRA et al., 2017).

The parameter that reflects water consumption by plants is crop evapotranspiration (ETc), which is of great relevance for irrigation and may have the same importance as rainfall, as it theoretically represents the amount of water needed to be replaced by irrigation (CAMARGO; SENTELHAS, 1997). ETc therefore reflects the maximum water loss of a crop at any stage of development under conditions of no water restriction, when there are optimum conditions of moisture and nutrients in the soil, so as to allow the potential production of the crop. Crop evapotranspiration expresses the simultaneous occurrence of the processes of soil water evaporation and plant transpiration and is a function of the energy balance, atmospheric demand and the supply of soil water to plants (CARVALHO et al., 2011; ADORIAN et al., 2015). Evapotranspiration varies between species and depending on crop management (ALLEN et al., 1998). Under natural conditions, the effective amount of water used by the plant is defined as actual evapotranspiration (ETr) (BISPO et al., 2017).

Actual evapotranspiration can be determined using the soil water balance method in a cropping system. This method quantifies, over a given period of time, the amount of water that has entered (precipitation, irrigation, capillary rise and surface runoff) or left (deep drainage, surface runoff and evapotranspiration) a control volume delimited by the soil surface and the depth of the root system (LIBARDI, 1995; PEREIRA et al., 2015a). In conjunction with the water balance method, gravimetry can be used to obtain soil moisture data for calculating the variation in soil water storage.

In cactaceous species, according to Morais et al. (2017), because they contain a high water content in their cladodes, the soil water balance method can cause erroneous interpretations of crop evapotranspiration, as it does not take into account the water content retained in the plants (in the parenchyma and vacuoles).

ETc can also be estimated by determining the reference evapotranspiration (ETo) and applying the crop coefficient (kc). This coefficient considers ideal conditions, without local limitations for crop development, or reduced evapotranspiration due to soil water restrictions, planting density, diseases, spontaneous vegetation, insects or salinity. The crop coefficient can be a single value (single kc) or the sum of two values. The double or "dual" crop coefficient approximation (dual kc) calculates the real increases in kc during each day as a function of the plant's development stages and the soil's surface moisture. In this way, the effects of transpiration and soil water evaporation on dual kc are determined separately as a function of the basal crop coefficient (kcb) and soil water evaporation coefficient (ke), respectively (SEDIYAMA, 1987; CARVALHO; OLIVEIRA, 2012; SILVA et al., 2012).

Due to the complexity of water transfer from the soil-plant-atmosphere system, there are various methods for estimating reference evapotranspiration (ETo), ranging from the simplest, empirically-based methods to those with a consistent physical basis. According to Ribeiro et al. (2016b), these methods include: Jensen-Haise (PEREIRA; VILLA NOVA; SEDIYAMA, 1997), Ivanov (JENSEN, 1973), Thornthwaite (THORNTHWAITE; MATHER, 1948), Priestley-Taylor (PRIESTLEY; TAYLOR, 1972) and Penman-Monteith (ALLEN et al., 1998). However, the Penman-Monteith method parameterised by the *Food and* Agriculture Organization of *the United Nations (*FAO) is considered the standard for estimating ETo. The method is based on physical aspects of the evaporation and transpiration processes and explicitly incorporates physiological and aerodynamic parameters of the reference crop (ALLEN et. al., 1998).

Pereira et al. (2017) when assessing the water balance in the soil cultivated with forage palm clones (IPA-Sertânia, Miúda and Orelha de Elefante Mexicana) in the Brazilian semi-arid region under the application of a fixed irrigation rate (7.50 mm) at intervals of 7, 14 and 28 days, for nine periods of 28 days, observed statistically significant differences between the components of the water balance in the soil throughout the crop cycle and between the irrigation conditions. However, there were no differences in water consumption between the clones with average daily real evapotranspiration values of 1.50 mm. The crop coefficient was also estimated by the same authors using the ratio between actual evapotranspiration and reference evapotranspiration, and an average value of 0.29 (±0.12) was obtained, with wide variation due to the seasonality of the water regime. The authors concluded that actual evapotranspiration was more dependent on the water regime than on atmospheric demand, and that evapotranspiration was intensified under irrigation frequencies of 7 days.

Silva et al. (2015b) obtained daily values of real evapotranspiration using the soil water balance method for three forage palm clones (IPA-Sertânia, Miúda and Orelha de Elefante Mexicana) grown under rainfed conditions, ranging from: 0.71 to 4.22 mm for IPA-Sertânia; 0.86 to 5.42 mm for Miúda and; 0.74 to 5.29 mm for Orelha de Elefante Mexicana.

Silva et al. (2014c), when assessing the real evapotranspiration of forage palm clones (Mexican Elephant Ear and Miúda) quantified using the residual soil water balance for 14-day intervals, using data on water content (9) and physical and water properties of the soil, found that the clones with the highest magnitudes of evapotranspiration showed greater efficiency in the use of the element sodium. This result may be indicative of the greater resistance of the *Opuntia* species in semi-arid environments, since the element sodium plays a major role in the stomata activity of CAM species, conferring greater efficiency in the use of water, which in the aforementioned study was obtained by the Mexican Elephant Ear, as evidenced by its greater productive performance in terms of fresh mass.

Water use efficiency is the ratio between atmospheric CO_2 uptake and daily water loss, which is a measure of the amount of water needed to produce photosynthetic products and therefore plant biomass. Therefore, water consumption by the plant is directly related to its development and production. Under conditions of adequate water availability, this efficiency is approximately 3 times higher in *Opuntia* than in C4 plants and around 5 times higher than in C3 plants, which is due to the relative amount of daily transpiration. The advantage may be even greater during drought, when the water stored in the succulent cladodes allows gas exchange to continue for a longer time compared to C3 and C4 plants, which have thin, non-succulent leaves (NOBEL, 1991; FAO, 1995). Gheyi et al. (2010) point out that succulence is also one of the mechanisms of adaptation to salt stress as a way of diluting salts in the cytosol; the volume of the cell increases, thus reducing the concentration of ions in the protoplast.

4. MATERIALS AND METHODS

4.1 Location and characterisation of the experimental area

The study was conducted under the open sky in the experimental area located on *campus* I of the Federal University of Campina Grande - UFCG, in the municipality of Campina Grande (7°12'52.56"S; 35°54'22.26"O and 532 m altitude), Paraíba, from 26 September 2017 to 11 December 2018, for a total of 442 days. According to the Koppen climate classification, the region's climate is of the Csa type, mesothermal, sub-humid, with a hot, dry dry season (4 to 5 months) and a rainy season from autumn to winter.

Throughout the experimental period, weather conditions were monitored by the automatic surface meteorological station of the National Meteorological Institute - INMET (7.22°S; 35.90°O and 546 m altitude) located approximately 1200 m away (in a straight line) from the experimental area (Figure 1).

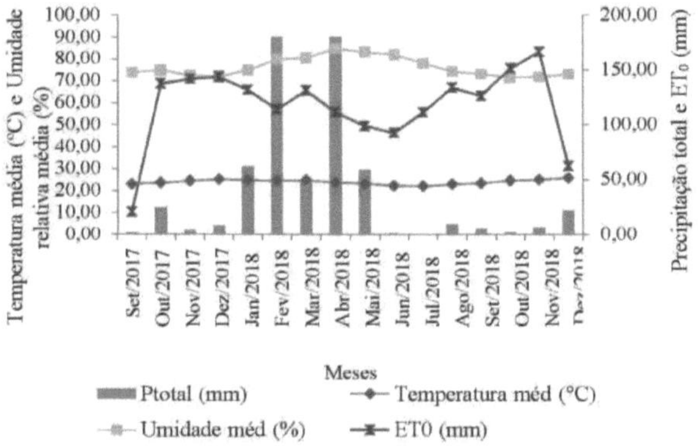

Figure 1. Average temperature (average temperature), average relative humidity (average humidity), total precipitation (Ptotal) and reference evapotranspiration (ET0) in the study region during the research period.

4.2 Experimental design and treatments

The experimental design adopted was a randomised block design (DBC) in a 4 x 4 factorial scheme with 4 repetitions, totalling 64 experimental plots. The factors consisted of four irrigation rates (L1 = 25%, L2 = 50%, L3 = 75% and L4 = 100%) as a function of soil water depletion from the total soil water capacity (TWC) and four levels of irrigation water salinity with electrical conductivity (ECa) of: S1 = 0.60 dS m^{-1} ; S2 = 3.00 dSm^{-1} ; S3 = 5.40 dS

m^{-1} and S4 = 7.80 dS m^{-1} ; applied to Elephant Ear palm *(Opuntia stricta (*Haworth) Haworth) (Figure 2). The control treatment was the one in which the irrigation rate replaced 100% of the total water capacity in the soil and with the lowest value of the electrical conductivity of the irrigation water (L4S1).

57 -L1S1B4	58 -L3S2B4	59 - L3S4B4	60 -L2S1B4	61 -L2S2B4	62 -L1S2B4	63 -L1S3B4	64 - L2S4B4
49 - L4S2B4	50 -L2S3B4	51 -L3S1B4	52 - L1S4B4	53 -L4S1B4	54 -L3S3B4	55 -L4S3B4	56 - L4S4B4
41 - L2S4B3	42 -L2S1B3	43 -L2S2B3	44 - L4S4B3	45 -L1S2B3	46 -L4S2B3	47 -L2S3B3	48 -L3S1B3
33 -L4S1B3	34 -L1S3B3	35 -L3S3B3	36 -L4S3B3	37 -L1S1B3	38 - L3S4B3	39 - L3S2B3	40 - L1S4B3
25 -L1S2B2	26 -L2S3B2	27 - L1S4B2	28-L1S3B2	29 -L4S2B2	30 -L3S2B2	31 -L2S1B2	32 -L4S3B2
17 -L3S1B2	18 - L4S4B2	19 - L3S4B2	20 -L2S2B2	21 -L3S3B2	22 -L1S1B2	23 - L2S4B2	24 -L4S1B2
9 - L3S3B1	10 -L1S1B1	11 -L2S3B1	12 - L1S4B1	13 -L2S1B1	14 -L4S3B1	15 -L4S2B1	16 - L3S4B1
1 - L2S2B1	2 - L3S1B1	3 - L4S1B1	4 - L3S2B1	5 - L2S4B1	6 - L4S4B1	7 -L1S2B1	8 -L1S3B1

Each colour represents the treatments with the same irrigation rate (L) as a function of the salinity levels of the irrigation water (S) and their respective blocks (B). The numbering that precedes the identification of the treatments corresponds to the number of the pot (1 to 64).

Figure 2 - Representation of the distribution of the experimental design in the research area

4.3 Preparation of experimental units and soil used

Preparing the area for the start of the research consisted of clearing the site using a small agricultural tractor and opening drains dug directly into the ground around the area to avoid possible flooding due to rainfall events. Periodically during the research, to maintain the area, the grass in the rows and rows between the pots was lowered using a brushcutter.

The research was carried out in 120-litre pots spaced 1.30 m apart between rows and 1.00 m between plants, with one plant per pot. The pots were painted white to encourage the reflection of solar radiation and prevent the soil from overheating. Two drains were installed at the bottom of each pot, connected by hoses to plastic collectors with a capacity of 2 litres (polyethylene terephthalate - PET - bottles) to allow drainage inside the pots. The pots were supported on two layers of bricks (Figure 3).

Figure 3: Initial preparation of the area for conducting the research

At the bottom, the pots were filled with gravel No. 1 (2.40 kg) covered with textile fabric, a layer of washed coarse sand (2.10 kg) and 170 kg of soil (0.268 m^3), according to the pot's capacity. The soil used to fill the pots came from Fazenda Cardeiro, in the rural area of the municipality of Soledade - PB. Soil analyses were carried out before the research began and at the end of the experimental period for its physical and chemical characterisation. The initial analysis was carried out by UFCG's Irrigation and Salinity Laboratory (LIS) and consisted of physical characterisation (Table 1), fertility (Table 2) and salinity (Tables 3A and 3B).

Table 1. Initial soil physical and water characterisation

Granulometry	Textural Classification	Soil density	Particle density	Porosity	Humidity (% dry soil basis)	
------- % --------- Sand Silt Clay	-	----------- g cm^{-3} ----------		%	0.10 atm	15.00 atm
82,79 11,86 5,35	White Sand	1,58	2,72	41,91	7,50	2,85

Table 2. Initial soil chemical characterisation (fertility)

Ca	Mg	In	K	SB	H	Al	CTC	M.O.	Assimilable P	pH H2O (1:2.5)
----- -		-meq	100 g^{-1}	soil	------	------		%	mg 100 g^{-1}	-
1,07	2,41	0,20	0,13	3,81	0,56	0,0	4,37	1,15	0,57	6,75

SB - Base sum; CTC - Cation exchange capacity; M.O - Organic matter.

Table 3A. Initial soil chemical characterisation (salinity)

EC	Chloride	Carbonate	Bicarbonate	Sulphate	Ca	Mg	K	In
mmhos cm^{-1}				..._meq L^{-1} __.	-			
0,53	5,00	0,00	0,22		1,50	3,50	3,76	2,56

Table 3B. Initial soil chemical characterisation (salinity)

PS	RAS	PST	Salinity	Soil class
%	-	-	-	-
22,17	1,62	4,57	Non-saline	Normal

PS - Percentage of saturation; RAS - Sodium adsorption ratio; PST - Percentage of exchangeable sodium.

The final soil analysis (Table 4) was carried out only for salinity by the Soil Laboratory of the Department of Soils and Rural Engineering at the Federal University of Paraíba (UFPB).

Table 4. Final soil chemical characterisation (salinity)

Sample	pH	EC	SO-24	Ca^{2+}	Mg^{2+}	In^{+}	K^{+}	COs'^{2}	HCO^{-}_{3}	Cl^{-}	RAS	PST
	-	dS m^{-1}			.... mmolcL^{-1}			-			mmolc 1 L^{-1}	%
L1S1	6,73	0,62	0,22	4,00	3,50	3,15	0,63	0,00	10,00	6,50	1,63	1,13
L1S2	7,50	0,92	0,09	7,00	10,50	5,81	0,42	0,00	12,50	14,00	1,96	1,61
L1S3	5,78	3,17	0,08	9,00	9,50	20,20	1,25	0,00	12,50	49,00	6,64	7,86
L1S4	5,23	4,88	0,11	16,50	14,50	35,59	2,41	0,00	12,50	69,00	9,04	10,77
L2S1	5,48	0,56	0,50	3,00	7,50	2,23	0,34	0,00	10,00	5,00	0,97	0,17
L2S2	5,27	1,09	1,04	0,50	5,00	7,32	0,67	0,00	10,00	12,50	4,41	4,99
L2S3	4,99	3,40	0,35	3,00	3,75	24,12	1,51	0,00	7,50	37,50	13,13	15,33
L2S4	4,98	3,96	0,51	3,50	11,50	25,65	1,81	0,00	12,50	40,00	9,37	11,16
L3S1	5,35	0,93	0,61	2,50	4,25	4,71	1,15	0,00	12,50	7,50	2,56	2,46
L3S2	5,19	1,39	0,39	0,75	6,00	9,06	1,06	0,00	5,00	10,00	4,93	5,67
L3S3	4,99	2,35	0,31	0,75	7,50	16,65	1,16	0,00	7,50	25,00	8,20	9,77
L3S4	5,32	2,66	0,32	2,25	5,25	15,27	1,34	0,00	10,00	22,50	7,88	9,39
L4S1	6,03	0,40	0,61	11,00	13,50	2,42	0,62	0,00	40,00	19,00	0,69	0,00
L4S2	5,81	1,93	0,08	5,00	6,75	10,66	1,26	0,00	12,50	31,50	4,40	4,97
L4S3	5,54	2,04	0,15	10,00	11,00	12,24	1,41	0,00	12,50	34,00	3,78	4,13
L4S4	7,55	2,59	1,45	7,50	15,00	15,71	1,22	0,00	7,50	31,50	4,68	5,35

ECes - electrical conductivity of the saturation extract; RAS - sodium adsorption ratio; PST - percentage of exchangeable sodium.

At the end of the research, the treatments: L1S3, L1S4, L2S3, L2S4, L3S3, L3S4 were classified as saline-sodic. Treatments L4S3 and L4S4 were classified as saline and the others as non-saline and non-sodic (Figure 4).

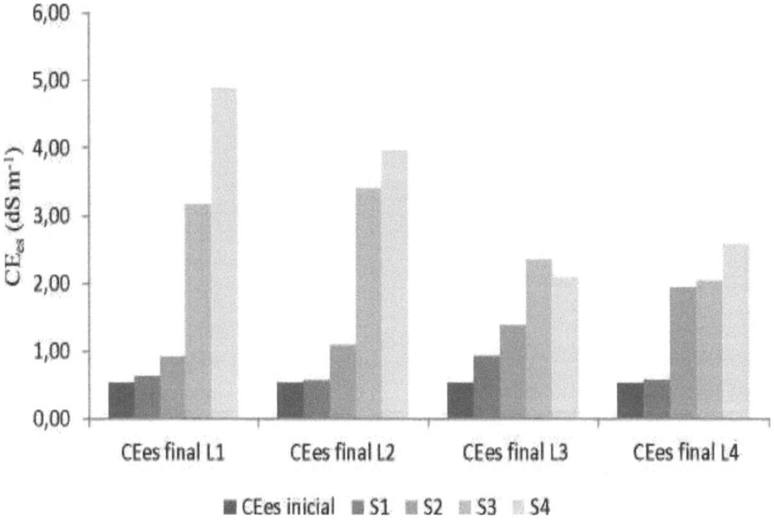

Figure 4: Initial and final electrical conductivity of the soil saturation extract (ECes) as a function of the averages per irrigation depth (L1 = 25; L2 = 50; L3 = 75 and; L4 = 100% of the total water capacity in the soil - CTA) and salinity levels of the irrigation water (S1 = 0.60; S2 = 3.00; S3 = 5.40 and S4 = 7.80 dS m^{-1}).

4.4 Plant material, fertilisation and cultivation

The cladodes of Mexican Elephant Ear palm *(Opuntia stricta* (Haworth) Haworth) used in the experiment came from the Lagoa Bonita Experimental Station of the National Semi-Arid Institute - INSA, located in the rural area of Campina Grande - PB. Secondary cladodes with approximate sizes were used for planting, endeavouring to maintain homogeneity between the plants. After the cladodes were cut at the Experimental Station, they were left in the shade for 15 days to lose excess humidity, allow the wounds caused by the cutting to heal and thus reduce the possibility of diseases. The cladodes were treated with Bordeaux mixture 48 hours before planting (Figure 5) to prevent infestation by fungi and bacteria (ROCHA, 2012; GAVA; LOPES, 2012). Planting was carried out at an angle of 45° to the ground so that the rackets would not topple over due to the wind and in an east-west direction.

Figure 5 Cladodes of Mexican Elephant Ear palm treated with Bordeaux mixture

Fertilisation was carried out according to the recommendations of Novais et al. (1991) for trials carried out in pots, maintaining the same doses for all the pots. Nitrogen (N), phosphorus (P) and potassium (K) were applied using urea, potassium chloride (KCl) and monoammonium phosphate (MAP), respectively. The total dose added to the soil was 18.84 g of urea, 50.85 g of KCl and 115.38 g of MAP. The fertiliser was divided into 30% as a foundation fertiliser and the difference was divided and applied monthly throughout the experimental period.

During the periods when scale insect incidences were identified, the fodder palm plants were sprayed with a solution containing water, neutral detergent and mineral oil, in the proportions recommended by Brito et al. (2008).

4.5 Irrigation

The treatments began 108 days after planting (11/01/2018) and were completed at the end of the cycle (442 days after planting), totalling 334 days of application, which was the time needed for the crop to establish itself and for all the plants to sprout at least one first-order cladode with an area greater than 18 cm^2 . To do this, the soil in the pots was brought to field capacity (09/01/2018) and as soon as the water content in the soil decreased (CTA), according to the pre-established percentages, the application of irrigation rates with saline water began (Figures 6 and 7 for soil at field capacity and sprouting of primary cladodes, respectively).

Figure 6. Adding rainwater to the soil to bring it up to field capacity

Figure 7. Emission of first-order cladodes of the Mexican Elephant Ear forage palm at the start of the treatments.

Irrigation was carried out manually using a graduated container with variable frequency. To assess moisture decay as a function of the percentage of total water capacity in the soil, soil samples were taken daily at depths of 15 and 30 cm (corresponding to the zone of greatest distribution of the root system of the *Opuntia stricta* (Haworth) Haworth forage palm) to determine moisture using the electric furnace method (MATOS et al., 2017). Soil was collected in the morning between 7am and 8am and irrigation was carried out in the late afternoon between 4.30pm and 5.30pm.

Based on the results of the physical analysis of the soil, the characteristic curve of the water in the soil was determined, adjusted by the van Genuchten model using the *Soil Water Retention Curve fit* (SWRT fit) computer programme, based on particle size (%) and particle density (g cm^{-3}) (Seki, 2007). In this way, the water content in the soil (volume-based moisture, cm^3 cm^{-3}) was determined on the basis of the moistures obtained by the gravimetric method using the electric oven, at the depth of the root system. The water content in the soil was used to define when to apply the irrigation rates (when to irrigate), the rate to be applied for the soil to return to field capacity (how much to irrigate) and to obtain soil water balance variables.

Initially, rainwater was used to bring the soil to field capacity before starting the treatments and to prepare the water with the different levels of electrical conductivity. The water was collected from the gutters of the vegetation houses located close to the experimental area and stored in 5000 litre polyethylene boxes on site. During the dry season, when the volume of precipitation was not enough to be used for irrigation, water from the Paraíba State Water and Sewage Company (CAGEPA) was used. The chemical analyses of the water were carried out by UFCG's Irrigation and Salinity Laboratory (Tables 5A and 5B).

Table 5A. Chemical parameters of the rainwater used in the research

pH	EC	Ca	Mg	In	K
-	$dS\ m^{-1}$	....	.... $mg\ L^{-1}$	-	
6,70	0,03	1,20	2,40	0,70	0,0

Chloride	Bicarbonate	Carbonate	Sulphate	Hardness	Alkalinity
	....	............-mg L⁻¹ -		-	
0,0	20,13	0,0	0,0	13,20	16,50

Table 5B. Chemical parameters of the water supply used in the research

pH	EC	Ca	Mg	In	K
-	$dS\ m^{-1}$		--------------------$mg\ L^{-1}$	-	
7,40	0,37	16,0	8,00	19,00	32,70

Chloride	Bicarbonate	Carbonate	Sulphate	Hardness	Alkalinity
	....		-$mg\ L^{-1}$ ___ -	....	-
46,79	75,64	0,0	35,04	73,12	62,00

The saline water was prepared for use in irrigation by adding commercial sodium chloride (without iodine), calcium chloride and magnesium chloride to the water in proportions of 7:2:1, respectively, in order to raise the electrical conductivity of the water, according to the methodology proposed by Richards (1954). The dilutions were made in four 500 litre polyethylene boxes, each box corresponding to a treatment with irrigation water salinity. The electrical conductivities of the water (ECa) of the treatments were checked weekly to correct them if necessary.

4.6 Reference evapotranspiration (ET0)

The automatic surface weather station of the National Meteorological Institute (INMET) was also used to obtain agrometeorological parameters. Using the data obtained from the station, reference evapotranspiration (ETo) was estimated using the Penman-Monteith - FAO 56 method, according to Allen et al. (1998):

$$ET_0 = \frac{0{,}408\Delta(R_n - G) + \gamma \frac{900}{T+273} u_2 (e_s - e_a)}{\Delta + \gamma(1 + 0{,}34 u_2)} \quad (1)$$

Where: ETo - daily reference evapotranspiration (mm d^{-1}); Rn - daily radiation balance (MJ m^{-2} d^{-1}); G - daily soil heat flux (MJ m^{-2} d^{-1}); T - average daily air temperature (° C); U2 - average daily wind speed at 2 m height (m s^{-1}); es - daily average water vapour saturation pressure (kPa); ea - daily average vapour pressure (kPa); A - slope of the vapour pressure curve at the T point (kPa °C^{-1}); y - psychrometric coefficient (kPa °C^{-1}).

Reference evapotranspiration (ETo) was estimated using the Penman-Monteith - FAO 56 method, using the computer programme Ref-ET: *Reference Evapotranspiration Calculator* (ALLEN, 1992).

4.7 Soil water balance

The actual evapotranspiration (ETr) of the forage palm *Opuntia stricta* (Haworth) Haworth was determined using the soil water balance method expressed by Libardi (1995) and adapted by Santana et al. (2016):

$$\Delta Arm = P + I \pm R \pm Q - ETr \quad (2)$$

Where: AArm - soil water storage variation (mm d^{-1}); P - precipitation (mm d^{-1}); I - irrigation (mm d^{-1}); R - surface runoff or run off (mm d^{-1}); Q - blade entering or leaving the lower contour (mm d^{-1}), which can be capillary rise (entry) or deep drainage (exit) and; ETr - real evapotranspiration (mm d^{-1}).

As the crop was grown in pots, surface run-off was considered to be zero, so Equation 2 was simplified into Equation 3:

$$ETr = P + I \pm Q - \Delta Arm \quad (3)$$

The soil water balance was carried out daily between 11/01/2018 and 06/12/2018, totalling 330 days, to determine monthly estimates of the real evapotranspiration of the Mexican Elephant Ear forage palm. The evapotranspiration estimates were determined in just one repetition of each treatment because of the need to remove soil daily at two depths in order to obtain moisture values using the gravimetric method. As the research was carried out in pots and for a crop with a long cycle, it wasn't feasible to carry out soil sampling in all the replications, as it was necessary to periodically switch between the replications of the same treatment so that the plant's roots wouldn't be exposed and thus jeopardise the plants' development.

4.7.1 Components of the soil water balance

Precipitation records were collected from the meteorological station in hourly values (mm h^{-1}), which were accumulated to obtain daily values (mm d^{-1}) for application in the soil water balance.

For calculation purposes, the same criterion used in the water balance developed by Thornthwaite and Mather (1955) and discussed by Pereira et al. (1997) was adopted. In this simplification, when the actual evapotranspiration (ET_r) was overestimated due to high rainfall values, the actual evapotranspiration was equalled to the reference evapotranspiration (ET_0).

The irrigation rates were determined based on the total water capacity of the soil (L1 = 25%, L2 = 50%, L3 = 75% and L4 = 100% of CTA), with variable irrigation frequency and determined by depleting the water content of the soil corresponding to the rates to be replenished. The total water capacity of the soil was determined from the total water availability of the soil, according to Salassier et al. (2009):

$$DTA = \frac{(CC - PM)}{10} \times d_a \quad (4)$$

$$CTA = DTA \times Z \quad (5)$$

Where: DTA - total soil water availability (mm cm^{-1} of soil); CC - field capacity (weight %); PM - wilting point (weight %); d_a - soil density (g cm^{-3}); CTA - total soil water capacity (mm) and; Z - effective root system depth (mm).

The variation in soil water storage was determined based on Lopes et al. (2004), for the 0.15 m depth:

$$\Delta Arm = (\theta_2 - \theta_1) \times z \quad (6)$$

Where: ΔArm - storage variation in the time interval considered (mm d^{-1});

θ_2 - average humidity at the end time (cm^3 cm^{-3}) corresponding to the humidity determined on the day; θ_1 - average humidity at the start time (cm^3 cm^{-3}) corresponding to the humidity of the previous day and; z - depth considered for the balance (mm).

The movement of water in the lower contour (Q) was determined by the Darcy-equation.

Buckingham (PREVEDELLO, 1996):

$$q = -K(\theta)\frac{d\psi t}{dx} \tag{7}$$

Where: q - soil water flow density (mm h^{-1}); K (0) - soil hydraulic conductivity (mm h^{-1}) and; $d\Psi_t/dx$ - total potential gradient at two depths in the control volume (Ψ_{t1}/ - Ψ_{t2}/ / xi - x2) corresponding to half the depth of the root zone and the end of the root zone (mm mm^{-1}).

The total soil water potential was estimated according to Gheyi et al. (2010):

$$\Psi_t = \Psi_m + \Psi_{os} + \Psi_g \tag{8}$$

Where Ψ m - matric potential (mm) Tos - osmotic potential (mm) e; Ψ g - gravitational potential (mm).

From the soil moisture values obtained by gravimetry and the soil water characteristic curve, the soil water matric potential (Tm) was determined. The osmotic potential was considered intrinsic to the matric potential, so Equation 8 was simplified into Equation 9:

$$\Psi_t = \Psi_m + \Psi_g \tag{9}$$

The gravitational potential was estimated according to Reichardt (1990):

$$\psi_g = ((d \times g \times z) \times 0{,}0010) \times 10 \tag{10}$$

Where: Ψ g - gravitational potential (mm); g - 1000 cm s^{-2} ; d - density of water (1 g cm^{-3}); z - height in relation to a reference, in this case corresponding to the depth from the soil surface to 15.0 cm (root zone); 0.0010 - unit conversion factor (barium to cm of H_2O) and; 10 - unit conversion factor (cm of H_2O to mm of H_2O).

The hydraulic conductivity of the unsaturated soil was estimated according to Prevedello (1996):

$$K(\theta) = K_0 w^L \left[1 - \left(1 - w^{\frac{1}{m}}\right)^m\right]^2 \tag{11}$$

Where: w - relative saturation and is determined by w = (θ - θ r)/(θ s - θ r); K0 - hydraulic conductivity of saturated soil (mm h^{-1}); L - empirical parameter and considered to be approximately 0.5 for most soils; 9 - current soil moisture (cm^3 cm^{-3}); θ r - residual soil moisture (cm^3 cm^{-3}); θ s - soil saturation moisture (cm^3 cm^{-3}) and; m - independent parameter estimated by the data adjusted from the van Genuchten model (1980), where m = 1 - 1/n (n - independent parameter estimated by the data adjusted from the van Genuchten model).

The hydraulic conductivity of saturated soil (K_0) was determined using the constant load permeameter method (YOUNGS, 1991).

4.8 Crop coefficient

The crop coefficient (k_c) was obtained according to Allen et al. (1998), at the end of the *Opuntia stricta* (Haworth) Haworth forage palm cycle, from the ratio between reference evapotranspiration (ET_o) and actual evapotranspiration (ET_r):

$$k_c = \frac{ETr}{ETo} \quad (12)$$

Where: kc - crop coefficient (dimensionless); ETr - actual evapotranspiration (mm d^{-1}); ETo - reference evapotranspiration (mm d^{-1}).

4.9 Growth variables

Growth assessments were carried out at the end of the cycle, 334 days after the start of the treatments or 442 days after planting, according to the methodology described by Borges et al. (2015) and consisted of: plant height (AP, cm), measured from the ground to the apex; plant width (LP, cm) measured considering the largest longitudinal ends of the plant; length of first-order or primary cladodes (CCP, cm) and second-order or secondary cladodes (CCS, cm); primary (LCP, cm) and secondary (LCS, cm) cladode width, primary (PCP, cm) and secondary (PCS, cm) cladode perimeter and primary (ECP, mm) and secondary (ECS, mm) cladode thickness. A tape measure was used to measure height, width, length and perimeter (Figures 8A, B and C) and a digital caliper with a precision of 0.05 mm was used to measure thickness.

(A) (B) (C)

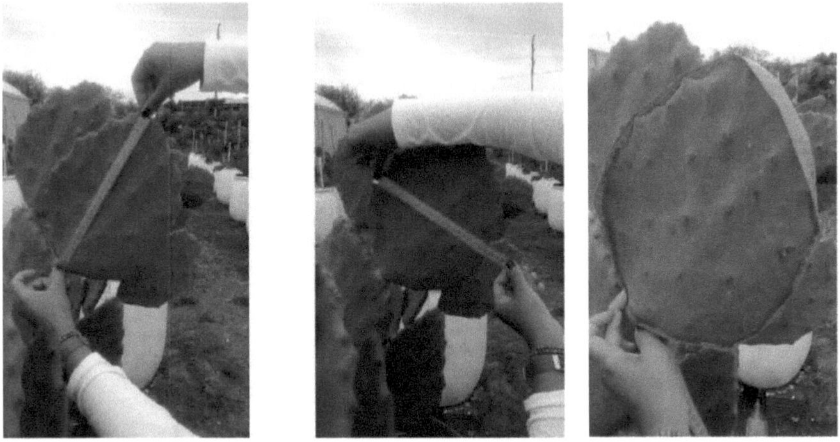

Figure 8. Growth assessments of cladode length (A), cladode width (B) and cladode perimeter (C).

The areas of the primary (ACP, cm^2) and secondary (ACS, cm^2) cladodes were estimated from the cladode length and width variables, using the methodology proposed by Santos et al. (2006) for fodder palm (*Opuntia):*

$$AC = CC * LC * 0.693 \qquad (13)$$

Where: AC - cladode area (cm^2); CC - cladode length (cm); LC - cladode width (cm) and; 0.693 - correction factor depending on the cladode.

Growth assessments were carried out on all the primary and secondary cladodes in the 64 experimental plots.

4.10 Physiological variables

The physiological assessments took place at the end of the cycle (442 days after planting or 334 days after the treatments were applied) and consisted of the relative water content (RWT, %) and electrolyte leakage (EE, %) variables.

Determining the relative water content was based on the methodology of Cairo (1995), while obtaining the percentage of electrolytes lost was based on the methodology described by Scotti & Thu Phan Thi (1997). However, both methodologies had to be adapted for the Mexican Elephant Ear forage palm because the plant is a succulent species and has its own morphological characteristics. In this case, instead of using circular leaf discs, three cylindrical samples extracted from primary and secondary cladodes of the same plant were used (Figures 9A and 9B). It was also necessary to increase the volume of deionised water added to the Baker from 20 mL to 30 mL, so that it would be sufficient to submerge the cylindrical samples.

(A) (B)

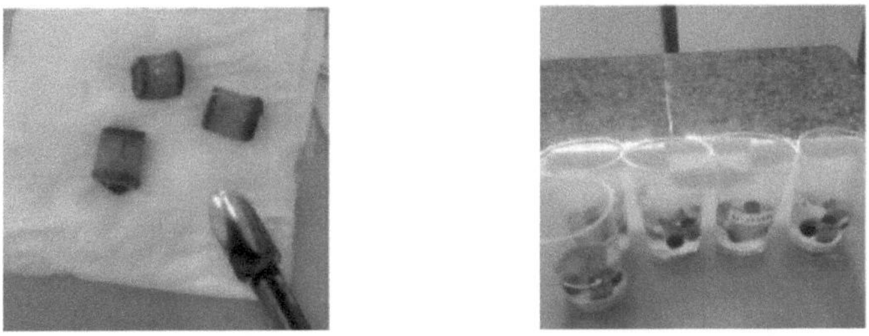

Figure 9. Adaptation of the proposed methodology for physiological evaluations with the extraction of cylindrical samples from Mexican Elephant Ear palm cladodes (A) and increase in the volume of deionised water (B).

The relative water content was found using Equation 14:

$$TRA = \frac{PF-PS}{PT-PS} \times 100 \quad (14)$$

Where: TRA - relative water content (%); PF - fresh matter weight (g); PS - dry matter weight (g); PT - turgid weight of the samples (g).

Electrolyte leakage in the cell membrane was obtained according to Equation 15:

$$EE = \frac{CEi}{CEf} \times 100 \quad (15)$$

Where EE - electrolyte leakage (%); CEi - initial electrical conductivity (dS m^{-1}); CEf - final electrical conductivity (dS m^{-1}).

4.11 Production variables

Production evaluations were carried out at 442 days after planting, equivalent to 334 days after the treatments were applied. The variables of number of primary (NCP, unit) and secondary (NCS, unit) cladodes and total number of cladodes per plant (NCT, unit) were obtained by direct counting, according to Borges et al. (2015).

The biomass production variables were obtained according to the methodology proposed by Silva et al. (2014b), also at the end of the experimental period. The cladodes of the Mexican Elephant Ear palm were cut, keeping only the main cladode in each experimental unit, to determine: primary cladode fresh mass (MFCP, kg per plant), secondary cladode fresh mass (MFCS, kg per plant) and total fresh mass per plant (MFT, kg per plant). Soon after the cladodes were removed, they were weighed on a digital scale with a precision of 0.05 g (Figures 10A and 10B).

(A) (B)

Figure 10. Cutting the fodder palm to assess production (A) and separating and identifying the cladodes for later weighing (B).

One unit of primary cladode and one unit of secondary cladode from each of the 64 experimental units were then separated, identified and weighed. These samples were used to determine the total dry mass (MST, kg per plant) using the ratio between the average fresh mass per plant (average between the fresh mass of the primary and secondary cladodes) and the average dry mass per plant (average between the dry mass of the primary and secondary cladodes):

$$MST = \frac{MFT \times MSM}{MFM} \quad (16)$$

Where: MST - total dry mass (kg per plant); MFT - total fresh mass (kg per plant); MSM - average dry mass (kg per plant) and; MFM - average fresh mass (kg per plant).

The samples were chopped up, placed in paper bags and dried in a forced-air oven at 65 °C, weighed every 48 hours until they reached a constant mass to determine the total dry mass.

4.12 Threshold salinity

The threshold salinity of the forage palm *(Opuntia stricta* (Haworth) Haworth) was determined using the first derivative of the equation for the graphical adjustment of the fresh mass production variables of primary and secondary cladodes and total fresh mass, as a function of the irrigation rates and salinity levels of the irrigation water.

4.13 Statistical analysis

The growth, physiology and production data were subjected to the distribution normality test (Shapiro-Wilk test) at the 5% probability level. After the normality test, analysis of variance was carried out using the F test at 1 and 5% probability levels for the individual factors of irrigation rates and salinity levels of the irrigation water and for the interaction between the factors (rates x salinity).

The data that showed significant effects of the treatments were adjusted using linear and quadratic polynomial regression (single factors and interaction between factors) and the response surface model (in the case of interaction between factors). The results of the interaction between the factors were presented in response surface format ($z = a + bx + cx^2 + dy + ey^2$) only for equations with graphical adjustments with a regression coefficient (R2) greater than 0.60. For lower R2 values in the response surface model, the interactions were represented by linear and quadratic polynomial regression.

The statistical analyses (Shapiro-Wilk and F-test) were carried out using Sisvar statistical *software*, version 5.6 (FERREIRA, 2011). The Table Curve 3D *software* (SILVA et al., 2019) was used to adjust the response surface models.

5. RESULTS AND DISCUSSION

5.1 Agrometeorology

5.1.1 Irrigation, reference evapotranspiration and actual evapotranspiration

The results of the average irrigation frequencies as a function of the treatments with irrigation water salinity levels are shown in Table 6.

Table 6. Average irrigation frequency as a function of irrigation rates and water salinity levels

Treatment	Frequency (days)	Treatment	Frequency (days)	Treatment	Frequency (days)	Treatment	Frequency (days)
L1S1	2	L2S1	5	L3S1	9	L4S1	10
L1S2	2	L2S2	6	L3S2	10	L4S2	10
L1S3	2	L2S3	5	L3S3	9	L4S3	13
L1S4	3	L2S4	6	L3S4	8	L4S4	12

L1S1, L1S2, L1S3 and L1S4 (25 per cent of total water capacity - TWC and 0.60; 3.00; 5.40 and; 7.80 dS m^{-1}, respectively); L2S1, L2S2, L2S3 and L2S4 (50 per cent of TWC and 0.60; 3.00; 5.40 and; 7.80 dS m^{-1}, respectively); L3S1, L3S2, L3S3 and L3S4 (75 per cent CTA and 0.60; 3.00; 5.40 and; 7.80 dS m^{-1}, respectively) and; L4S1, L4S2, L4S3 and L4S4 (100 per cent CTA and 0.60; 3.00; 5.40 and; 7.80 dS m^{-1}, respectively).

According to the results obtained, the average irrigation frequencies varied between 2 and 13 days, depending on the treatments. In general, the lowest frequencies were obtained for the highest saline levels of irrigation water (S3 = 5.40 dS m^{-1} and S4 = 7.80 dS m^{-1}).

The lower frequencies for the irrigation water with the highest saline concentrations occurred because the salts present in the water and deposited in the soil caused changes in its physical and chemical properties. This process could be observed with the increase in the electrical conductivity of the soil saturation extract (EC_{es}) for these treatments (Table 4 presented in the methodology).

According to Queiroz et al. (2010), the accumulation of salts in the soil due to irrigation compromises the soil's chemical, physical and biological properties. Especially in semi-arid climate regions, where evaporation predominates over precipitation. Salinisation of the soil reduces the osmotic potential of its solution and can lead to a reduction in the availability of water for plants (SOUZA et al., 2019).

When comparing the same salt level for the different water levels applied, there was a progressive reduction in irrigation frequencies. This is because the timing of the irrigation application was determined by the decay of water in the soil, as a function of the total water capacity in the soil, through the water balance. Therefore, as the water levels increased, a longer period (lower irrigation frequency) was needed for the water in the soil to decay and for irrigation to be applied.

The results of rainfall (mm), reference evapotranspiration (ETo, mm), irrigation rates (mm) and actual evapotranspiration (ETr, mm) are shown in Figure 11 (A, B, C and D) for each treatment considered of irrigation rates and levels of irrigation water salinity.

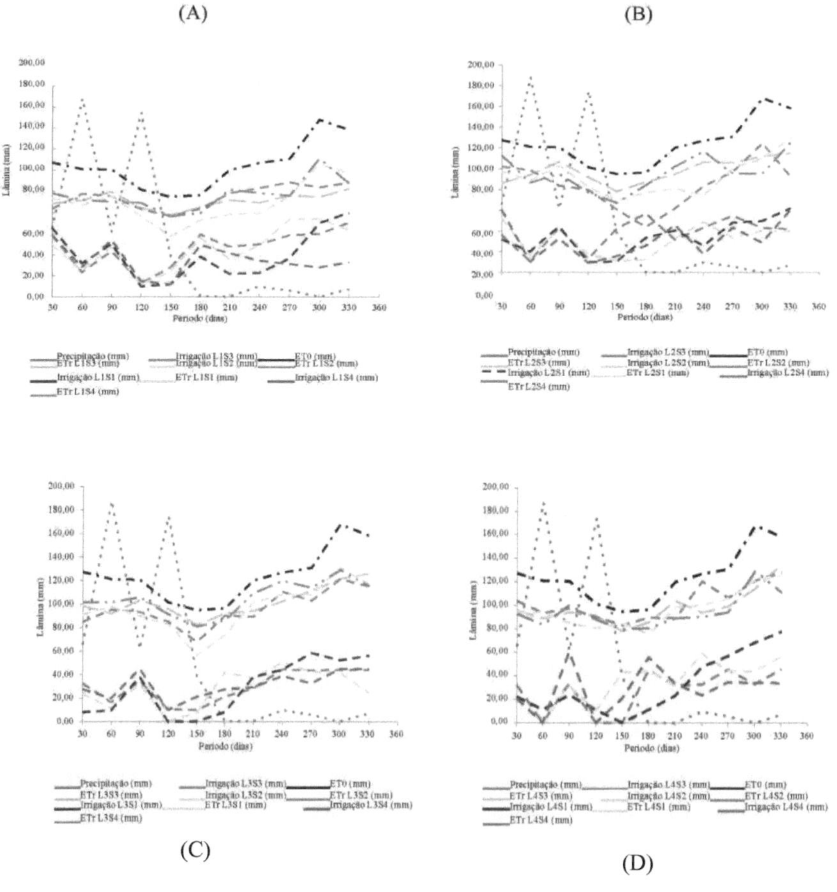

L1S1, L1S2, L1S3 and L1S4 (25% of total water capacity - TWC and 0.60; 3.00; 5.40 and; 7.80 $dS\ m^{-1}$, respectively); L2S1, L2S2, L2S3 and L2S4 (50% of TWC and 0.60; 3.00; 5.40 and; 7.80 $dS\ m^{-1}$, respectively); L3S1, L3S2, L3S3 and L3S4 (75% of CTA and 0.60; 3.00; 5.40 and; 7.80 $dS\ m^{-1}$, respectively) and; L4S1, L4S2, L4S3 and L4S4 (100% of CTA and 0.60; 3.00; 5.40 and; 7.80 $dS\ m^{-1}$, respectively).

Figure 11. Monthly soil water balance corresponding to the period from 11/01/2018 to 06/12/2018 (330 days) as a function of irrigation rates and water salinity levels.

In general, the soil water balance showed the same trend for irrigation up to 120 days for all treatments (Figure 11). This same period saw the highest rainfall values, totalling 489.00 mm, which corresponded to 88.94% of the total rainfall (after the treatments were applied). It is possible that the rainfall de-characterised the irrigation rates applied in this interval, as the rainfall recharge to the soil was the same for all treatments, returning the soil to values close to field capacity. The rainfall may also have favoured the leaching of salts deposited by the irrigation water into the sandy loam soil. As the intensity of the rainfall decreased after 120 days, there was variability in the irrigations within the same blade for different salinity levels.

Fonseca et al. (2019) in a study carried out with different irrigation rates and frequencies, water salinity levels and Giant fodder palm *(Opuntia ficus-indica* Mill), stated that after the rainy season there was no significant difference in soil salinity between the different water application conditions. The same was observed in the current study during the period of greatest rainfall.

The accumulated reference evapotranspiration (ETO) for the 330 days evaluated was 1364.37 mm. There was a decline in ET0 values in the interval between 90 and 180 days (April to July) after the treatments were applied, due to the region's climatic characteristics for this period (Figure 11). The maximum monthly demand forETo (167.74 mm) within the period of evaluation of the agrometeorological variables occurred at 300 days (between late October and early November), while the minimum monthly demand (94.67 mm) was observed at 150 days (late May and early June).

For the L1 blade (25 % of total water capacity - CTA), the total irrigation values increased up to the salinity of 5.40 dS m^{-1} (S3) and decreased as the salinity increased to the level of 7.80 dS m^{-1} , with this level showing the lowest irrigation blade (Figure 11A). The total water applied according to the different levels of irrigation water salinity for L1 was: 434.88 mm (S1 = 0.60 dS m^{-1}); 505.37 mm (S2 = 3.00 dS m^{-1}), 507.63 mm (S3 = 5.40 dS m^{-1}) and 380.67 mm (S4 = 7.80 dS m^{-1}). Despite this, treatments L1S2 and L1S3 showed similar values. The percentage difference between the maximum and minimum irrigated values within the same blade for the different electrical conductivities of the water was 25.01%. This demonstrated the influence of the accumulation of salts in the irrigation water on the soil and its effect on the fodder palm crop. Another result that reinforces the effect of salinity was shown by the frequency of irrigation as a function of saline levels, which was approximately 2 days for S1, S2 and S3 and 3 days for S4. In addition, the final electrical conductivity of the soil saturation extract was higher for treatment S4 (Table 4 presented in the methodology). With salinisation, the soil's osmotic potential tends to remain more negative than the plant's potential and so the plant is unable to absorb the water present in the soil since water moves from the highest potential to the lowest potential. For absorption to take place in these cases, the plant needs to carry out osmotic adjustment, a way of making its osmotic potential more negative than the soil's and thus be able to absorb water. In relation to the actual evapotranspiration (ETr) of the treatments with a blade of 25% of the CTA, the highest magnitudes were obtained for the L1S4 treatment with a total accumulated ETr equal to 1057.11 mm and the minimum total accumulated ETr for L1S1 equal to 976.24 mm.

Although L1S4 had the lowest irrigation rate, this treatment had the highest ETr compared to the other salt levels. This is possibly due to the fact that the sodium ion (Na^+) can partly replace potassium (K^+) as an osmotically active solute. This ion thus participates in osmoregulation and favours a potential gradient. Stomatal opening occurs when solutes are actively accumulated in the guard cells. This accumulation (and consequent reduction in water potential in the guard cells) causes an osmotic movement of water into the guard cells, increasing their turgor pressure. Stomatal closure occurs through the reverse process, i.e. a decrease in the concentration of solutes in these cells. In addition, the element sodium is essential for the regeneration of phosphoenolpyruvate, the substrate for the first carboxylation in the CAM pathway, and contributes to the Calvin cycle not being interrupted. Most CAM species require sodium ions (WILLADINO; CAMARA, 2010; TAIZ; ZEIGER, 2013).

Although forage palm is not considered a salinity-tolerant plant, Willadino & Camara (2010) stated that among the effects of Na+ accumulation in the plant, the reduction in K+ concentration as a result of the increase in salinity stands out, which is recognised as a tolerance mechanism of some plants associated with the capacity for selective absorption of K^+.

The participation of sodium in stomata regulation was also observed by Silva et al. (2014c). The authors found for fodder palm clones, particularly the species of the *Opuntia* genus, that higher magnitudes of ETr showed greater efficiency in the use of sodium. The authors found an average accumulated ETr of 1,418 mm per cycle, a result higher than that found in the present study for the L1 blade (25% of CTA) with the highest electrical conductivity of the irrigation water (L1S4 - 1,057.11 mm).

Slope L2 (50 per cent of CTA) showed little variation in the accumulated irrigation values between the saline levels for the 330 days after the treatments began (Figure 11B). The results were: 375.42 mm (L2S1); 344.51 mm (L2S2); 415.31 mm (L2S3) and; 347.39 mm (L2S4). The maximum total irrigation occurred for treatment (L2S3). The same irrigation frequency was observed for L2S1 and L2S3, of approximately 5 days, and 6 days for L2S2 and L2S4 throughout the period evaluated. The highest irrigation rates had the lowest irrigation frequencies. With regard to ETr, the highest magnitudes were found for salinity treatments S3 (5.40 dS m^{-1}) and S4 (7.80 dS m^{-1}) with accumulated values of 1076.39 mm and 1056.91 mm, respectively. The other salinities showed ETr of 1012.98 mm (S1 = 0.60 dS m^{-1}) and 926.65 mm (S2 = 3.00 dS m^{-1}). This result corroborates Silva et al. (2014c) and Taiz & Zeiger (2013), whose authors affirmed the participation of sodium in stomata regulation, thus contributing to an increase in transpiration rate.

According to Basu et al. (2017), excess salts in plant roots alter the transpiration process, which is a crucial step in regulating and maintaining water in the plant. However, according to Anower et al. (2017), some plants already naturally show a degree of tolerance to salts and this is due to the development of various mechanisms ranging from morphological changes to physiological and biochemical modifications. In the case of this research, there were results indicating that the Mexican Elephant Ear forage palm showed a degree of tolerance to salts in the irrigation water, possibly caused by osmoregulation.

As for the results obtained for the L3 slope (75% of the CTA) at the different levels of irrigation water salinity (Figure 11C), the L3S4 treatment (7.80 dS m^{-1}) had the highest accumulated irrigated value with 337.42 mm, followed by S3 (5.40 dS m^{-1}) with 326.78 mm. Treatments L3S1 (0.60 dS m^{-1}) and L3S2 (3.00 dS m^{-1}) had values of 310.32 mm and 309.64 mm, respectively. The irrigation frequencies were 9, 10, 9 and 8 days, corresponding to the saline levels of the irrigation water in ascending order. Unlike the L1 and L2 treatments, the L3 treatment had the highest frequency at salinity S4. This salinity also saw greater water consumption by the crop compared to the other L3 treatments, whose maximum accumulated ETr was 1153.44 mm. However, a similar value was found for L3S3 with an ETr of 1116.88 mm. The L3S1 and L3S2 treatments had practically the same ETr with 1054.83 mm and 1054.60, in that order.

Fonseca et al. (2019) studied the Giant fodder palm *(Opuntia ficus-indica* Mill) under irrigation rates as a function of reference evapotranspiration and different irrigation frequencies (rainfed; 5 litres of water per linear metre every 15 days; 7% of ET0 with a 15-day irrigation interval; 15% of ET0 with a 7-day irrigation interval; 33% of ET0 with a 3-day irrigation interval; 50% of ET0 with a 2-day irrigation interval; and 100% of ET0 with daily irrigation), with irrigation water salinity of 3.60 dS m^{-1} . These authors found that even with the use of saline water, irrigation offers better conditions for crop development, perhaps due to the increase in photosynthetic rates.

The L4 (100 per cent CTA) treatments had the lowest irrigation frequencies, in general, compared to the other irrigated slopes. The frequencies were 10 days for L4S1 (0.60 dS m^{-1}) and L4S2 (3.00 dS m^{-1}), 13 days for L4S3 (5.40 dS m^{-1}) and 12 days for L4S4 (7.80 dS m^{-1}). The irrigation rates were 352.76 mm, 385.57 mm, 299.22 mm and 317.95 mm for S1, S2, S3 and S4, respectively. The results (Figure 11D) showed that the greatest accumulation of irrigation occurred for the L4S2 treatment and, for this same treatment, the greatest magnitude of ETr was observed (1124.13 mm). There was an increase in the ETr value up to the S2 saline level and, as the salinity of the irrigation water increased, there was a decrease in the actual evapotranspiration values. Thus, the L4 treatment differed from the other levels in that the highest ETr values were found for the highest level of irrigation water salinity (7.80 dS m^{-1}). The lower frequency of irrigation shown by the combinations of treatments with L4 may have influenced this result.

When evaluating the same level of salinity of the irrigation water for the different blades, it was found that as the blades increased, the irrigation frequencies decreased and that the final electrical conductivity of the soil saturation extract (EC_{es}) was a function of the irrigation frequency and the saline levels of the water. The treatments that promoted the highest magnitudes of ETr as a function of varying saline levels for the same blade were: S4 (7.80 dS m^{-1}) for blade L1 (25% of CTA); S3 (5.40 dS m^{-1}) for blade L2 (50% of CTA); S4 (7.80 dS

m^{-1}) for blade L3 (75% of CTA) and; S2 (3.00 dS m^{-1}) for blade L4 (100% of CTA). The treatments were: L3S3; L3S4; L4S2 and; L4S1. It was therefore found that the treatments that coincidentally showed the maximum accumulated ETr, both for the different water levels at the same salinity level and vice versa, were the L3S4 and L4S2 combinations, with L3S4 standing out, as shown in Table 7.

Table 7. Accumulated real evapotranspiration (ETr) of Mexican Elephant Ear forage palm from 11/01/2018 to 06/12/2018 as a function of irrigation treatments and water salinity levels.

Treatment	ETr (mm)	Treatment	ETr (mm)	Treatment	ETr (mm)	Treatment	ETr (mm)
L1S1	976,24	L2S1	1012,98	L3S1	1054,83	L4S1	1067,27
L1S2	1040,05	L2S2	926,65	L3S2	1054,60	L4S2	1124,13
L1S3	999,31	L2S3	1076,39	L3S3	1116,88	L4S3	1079,36
L1S4	1057,10	L2S4	1056,91	L3S4	1153,44	L4S4	1039,44

L1S1, L1S2, L1S3 and L1S4 (25 per cent of total water capacity - TWC and 0.60; 3.00; 5.40 and; 7.80 dS m^{-1} , respectively); L2S1, L2S2, L2S3 and L2S4 (50 per cent of TWC and 0.60; 3.00; 5.40 and; 7.80 dS m^{-1} , respectively); L3S1, L3S2, L3S3 and L3S4 (75 per cent CTA and 0.60; 3.00; 5.40 and; 7.80 dS m^{-1} , respectively) and; L4S1, L4S2, L4S3 and L4S4 (100 per cent CTA and 0.60; 3.00; 5.40 and; 7.80 dS m^{-1} , respectively).

Divincula et al. (2019), when assessing the water requirements of the forage palm crop *(Opuntia ficus-indica* Mill) in the agreste region of Alagoas, obtained an accumulated crop evapotranspiration of 637.84 mm over the course of the cycle. Crop evapotranspiration was determined by the authors using a drainage lysimeter. This result was lower than the ETr values found for the different treatments in this study; however, the cycle considered by Divincula et al. (2019) was 152 days, while in this study the agrometeorological assessments were carried out over 330 days.

Silva et al. (2014c) obtained real evapotranspiration accumulated by the soil water balance method for fodder palm *(Opuntia),* during a 499-day cycle, of 1418 mm with a daily average of 2.84 mm. This result was higher than that found in the present study, which was also obtained using the soil water balance method, but for an evaluated period of 330 days and under irrigated conditions.

Jardim et al. (2017), in an experiment conducted in the municipality of Serra Talhada in Pernambuco with irrigated forage palm *Opuntia stricta* (Haworth) Haworth based on the 75 per cent fraction of reference evapotranspiration, obtained accumulated ETr of 645 mm. This result was obtained for the fourth crop cycle. The authors also found that ETr differed between production cycles. To determine the agrometeorological variables, the authors also used the soil water balance methodology.

Queiroz et al. (2016), when using the soil water balance to determine the accumulated ETr of the Mexican Elephant Ear forage palm under different water levels in Brazilian semi-arid conditions, obtained an accumulated ETr of 985.77 mm for the 380-day cycle. The authors found that Mexican Elephant Ear, under conditions of greater water availability in the soil, does not fully utilise it in the evapotranspiration process. This result was lower than that obtained in this study for a shorter evaluation period than that of the aforementioned authors. It is inferred that the salinity of the irrigation water favoured an increase in the crop's real evapotranspiration.

Authors such as Nobel (1991) and Freire (2012) have stated that forage palm is sensitive to salinity and is therefore not considered a halophyte plant. However, according to Willadino and Camara (2010), in halophytes, salinity tolerance is often positively correlated with Na^+ concentration. The authors also stated that when related to high cytoplasmic concentrations of K^+, Na^+ favours the plant's osmotic adjustment.

In addition to sodium, the water used to irrigate the fodder palm in this study contained the elements calcium and magnesium in the forms of calcium chloride and magnesium chloride. According to Taiz & Zeiger (2013), magnesium is an important constituent of the structure of the chlorophyll molecule and plays a specific role in the activation of photosynthesis and respiration enzymes. The increase in its absorption by the plant, related to the concentration of the elements calcium and potassium, can promote an increase in crop productivity, although this varies between species. In this way, the other elements present in the irrigation water may also have favoured an increase in actual evapotranspiration.

According to Franco-Salazar & Véliz (2008), increasing salinity could induce osmotic adjustment or osmoprotection of the photosynthesising apparatus in fodder palm. This would explain why, in this study, the maximum ETr was observed at the highest level of irrigation water salinity (L3S4 - 1153.44 mm).

The higher evapotranspirometric rate is a physiologically important mechanism for the plant because it reduces heat stress, as well as favouring the process of nutrient absorption through a potential gradient.

The results of the accumulated average irrigation (mm) and actual evapotranspiration (ETr, mm) as a function of blade treatments only (average per blade), regardless of the salinity of the irrigation water, are shown in Figure 12.

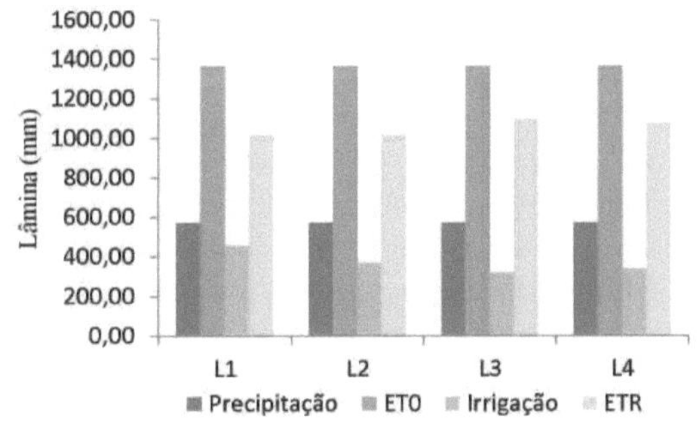

L1; L2; L3 and L4 correspond to 25, 50, 75 and 100 per cent of the total soil water capacity (TWC), respectively.

Figure 12. Accumulated values of precipitation, reference evapotranspiration (ETO), irrigation and actual evapotranspiration (ETr), regardless of the electrical conductivity of the water (average per blade), in the period from 11/01/2018 to 06/12/2018 (330 days).

Based on the results obtained, it was found that the highest average irrigation rates were observed for L1 (25% of total water capacity - TWC), with an average irrigation frequency of approximately 2 days. This value corresponded to the accumulated average of the irrigations applied at all saline levels for the corresponding slope, totalling 457.12 mm. The other levels (L2 - 50%, L3 - 75%, L4 - 100% of CTA) showed little variation between them with average irrigated values of 370.66 mm, 321.04 mm and 338.87 mm, respectively. The average irrigation frequencies as a function of the water levels, regardless of water salinity, were approximately: 6 days (L2), 9 days (L3) and 11 days (L4). As the percentage of total water capacity in the soil increased to define when to irrigate, irrigation frequencies decreased. However, this was not a determining factor in the gradual reduction in the amount of water applied between the different rates. Similarly, the greater availability of water through irrigation (457.12 mm) did not result in maximising the crop's real evapotranspiration. The average real evapotranspiration per blade was: 1018.18 mm (25% of CTA - L1); 1018.23 mm (50% of CTA - L2); 1094.94 mm (75% of CTA - L3) and; 1077.55 mm (100% of CTA - L4).

Campos (2018) found that the use of irrigation increases the actual evapotranspiration of fodder palm compared to rainfed cultivation, due to the increased availability of water to the plant. The greater availability of water also increases the evaporation of water in the soil.

Although the L3 blade (75% of the CTA) had the lowest average irrigation value, this blade provided the highest magnitude of ETr (1094.94 mm) compared to the others. And when evaluated separately in relation to salinity levels, it was observed that this same blade showed the highest ETr for salinity S4 (7.80 dS m^{-1}), as discussed above. This confirmed the effect of irrigation water salinity on the irrigated water levels and on the actual evapotranspiration of Mexican Elephant Ear palm. It was also found that the main influences on agrometeorological variables came from salinity levels and not from irrigation rates. Sodium may have played a fundamental role in the results found, favouring the plant's osmotic regulation and, consequently, an increase in transpiration rates. Another factor that should be considered is the evaporation of water from the soil. According to Gheyi et al. (2010), soils exposed to a higher concentration of sodium in the exchangeable phase show an increase in retained moisture, due to the modification of their structure by sodicity and/or the absence of ions in solution.

5.1.2 Crop coefficient

The results of the estimated crop coefficients (kc) for Mexican Elephant Ear palm are shown in Table 8.

Table 8. Crop coefficients (kc) of Mexican Elephant Ear palm obtained at 442 days into the cycle

Treatment	kc	Treatment	kc	Treatment	kc	Treatmen	kc
L1S1	0,71	L2S1	0,75	L3S1	0,77	L4S1	0,79
L1S2	0,77	L2S2	0,68	L3S2	0,78	L4S2	0,83
L1S3	0,75	L2S3	0,80	L3S3	0,83	L4S3	0,80
L1S4	0,78	L2S4	0,78	L3S4	0,85	L4S4	0,77
Average 0.		75-0	.	75-0	.	81-0	.80

L1S1, L1S2, L1S3 and L1S4 (25 per cent of total water capacity - TWC and 0.60; 3.00; 5.40 and; 7.80 dS m^{-1} , respectively); L2S1, L2S2, L2S3 and L2S4 (50 per cent of TWC and 0.60; 3.00; 5.40 and; 7.80 dS m^{-1} , respectively); L3S1, L3S2, L3S3 and L3S4 (75 per cent CTA and 0.60; 3.00; 5.40 and; 7.80 dS m^{-1} , respectively) and; L4S1, L4S2, L4S3 and L4S4 (100 per cent CTA and 0.60; 3.00; 5.40 and; 7.80 dS m^{-1} , respectively).

According to the results, the maximum kc value (0.85) found for Mexican Elephant Ear was provided by treatment L3S4 (75 per cent CTA and 5.40 dS m^{-1}). The minimum kc value (0.68) was observed for the L2S2 treatment (50% CTA and 3.00 dS m^{-1}).

In relation to the kc values for the same irrigation rate and different water salinity levels, treatments L3S4 (kc = 0.85), L4S2 (kc = 0.83), L2S3 (kc = 0.80) and L1S4 (kc = 0.78) had the highest crop coefficients. And for the same saline level with the variation in slats, the highest values were found for: L3S4 (kc = 0.85), L3S3 (kc = 0.83), L4S2 (kc = 0.83) and L4S1 (kc = 0.79). The L3S4 and L4S2 treatments stand out in terms of both the water levels and the salinity levels of the irrigation water. This is because the crop coefficient depends on the ratio between the crop's actual evapotranspiration and the reference evapotranspiration and, for these treatments, the maximum actual evapotranspirations were also observed.

Determining the crop coefficient in a given climate, soil, plant and water condition should be considered representative, since this coefficient is one of the most efficient in representing the biophysical phenomenon of evapotranspiration (SILVA et al., 2017b).

Divincula et al. (2019) found average kc of 0.72, 0.84 and 0.48 for forage palm *(Opuntia ficus-indica* L.) in the conditions of the agreste region of Alagoas, using the Penman-Monteith, Hargreaves-Samani and FAO-Radiation methods, respectively. The values were similar to those obtained in this study, with the exception of the kc determined by the FAO-Radiation method. The authors kept the soil at field capacity throughout the experimental period.

Jardim et al. (2017) obtained an average kc equal to 0.57 for the same forage palm species researched *(Opuntia stricta* (Haworth) Haworth). The authors also determined ET_r using the residual from the soil water balance method under irrigated conditions, in order to subsequently determine the crop coefficient. However, the kc value was around 16% lower than the lowest value found in the current study. This difference may be due to the ETr value estimated by the authors, which was lower than that observed in this research, which influenced the difference between the crop coefficients.

Santos et al. (2015) determined a higher average crop coefficient than this study, with kc equal to 1.54 using the Penman-Monteith method and 1.23 using the Hargreaves-Samani method. The work was carried out with fodder palm subjected to irrigation shifts, however, the authors did not inform the genus of palm used in the research. Crop coefficients above 1.00 indicate that the plant has an evapotranspirometric demand greater than the reference evapotranspiration. The kc values vary according to the type of crop, its stage of development, the length of its vegetative cycle and local climatic conditions (BERNARDO; MANTOVANI, 2006).

Queiroz et al. (2016) obtained an average crop coefficient of 0.52 for the Mexican Elephant Ear palm, with little variation over time. The highest observed kc value occurred at the end of the cycle, due to maximum vegetative growth. The authors concluded that these changes in kc throughout the cycle were mostly due to the crop's physiological adjustments to the environment.

With regard to the results of the average crop coefficients per irrigation blade, it was observed that the L3 blade (75% of the CTA) had a kc equal to 0.81 (Table 8), which was the maximum value. This same blade also had the highest magnitude of ETr (1094.94 mm).

According to Bernardo & Mantovani (2006), the higher the local evapotranspirometric demand or the more sensitive the plant is to soil water deficit, the higher the kc value will be.

According to Lopes et al. (2011), kc is a parameter related to the environmental and physiological factors of plants.

In general, the agrometeorological variables assessed may have been influenced not only by the local climate and growing conditions, but also by the availability of water to the plant through irrigation. Under irrigated conditions, certain CAM plant species have the ability to adjust their metabolism to C3 (SILVA et al., 2015b). However, it was observed that the main influence on the results was due to the salinity of the irrigation water, which may have favoured the plant's osmotic adjustment.

In the case of the Mexican Elephant Ear forage palm, Queiroz et al. (2016) concluded that its vegetative organs are highly adaptable to variable soil and climate conditions.

Therefore, the kc describes the phenological and physiological development of a particular crop in relation to the reference evapotranspiration and represents the water consumption of a specific crop which is of relevant importance for estimating its water requirement. The water requirement is, in turn, necessary for dimensioning irrigation systems and operationalising irrigated perimeters (CLARK et al., 1996). In this way, the kc associated with the reference evapotranspiration for determining the evapotranspiration of the Mexican Elephant Ear forage palm makes it possible to carry out efficient irrigation management, which results in the optimised use of water.

5.2 Growth

5.2.1 Primary cladode

The results of the analysis of variance using the F test at the 1 and 5 per cent probability level for the variables length, width, area, perimeter and thickness of the primary cladode at 334 days after the treatments were applied, corresponding to a total cycle of 442 days, are shown in Table 9.

Table 9. Summaries of the analysis of variance for primary cladode length (PCL), primary cladode width (PCW), primary cladode area (PCA), primary cladode perimeter (PCP) and primary cladode thickness (PCT) of Mexican Elephant Ear forage palm at 442 days of the cycle.

Source of variation	GL	Mean squares (QM)				
		CCP[1]	LCP	ACP[1]	PCP	ECP
Blade	3	66,51^{ns}	6,46^{ns}	44623,12^{ns}	48,01^{ns}	1,45^{ns}
Salinity	3	41,13^{ns}	5,83^{ns}	19373,73^{ns}	51,82^{ns}	3,67^{ns}
Blade x Salinity	9	50,73^{ns}	1,71^{ns}	17987,23^{ns}	18,07^{ns}	2,56^{ns}
Block	3	71,47^{ns}	8,04^{ns}	53399,78^{ns}	103,92^{ns}	0,45^{ns}
Error	45	55,33	4,43	29940,33	49,86	2,08
CV (%)	-	9,80	8,68	12,83	8,91	9,28
Overall average	-	30,02	24,27	551,86	79,25	15,56

^{ns} - not significant by F test (p>0.05); [1] data transformed by square root.

According to the data obtained (Table 9), there were no statistically significant differences ($p>0.05$) for any of the growth variables analysed. It was therefore found that the treatments applied had no influence on the growth parameters of the primary cladode.

Average values between treatments were observed for primary cladode length (CCP), width (LCP), area (ACP), perimeter (PCP) and thickness (ECP), respectively: 30.02 cm; 24.27 cm; 551.86 cm^2 ; 79.25 cm and; 15.56 mm.

Although Donato et al. (2014) state that the morphometric characteristics of fodder palm are little influenced by management, which was also verified in this study, when comparing them with those of other authors, it was observed that the results obtained differ from studies such as those by Silva et al. (2015c) and Lima et al. (2016). These authors found that the adoption of forage palm management practices had significant effects on plant growth, whether through the use of irrigation, different planting spacings or the provision of fertiliser.

Silva et al. (2015c), when assessing the growth of forage palm clones under Brazilian semi-arid conditions and their relationship with meteorological variables, found statistically significant differences for the average length of the primary cladode, with a value of 27.73 cm for Orelha de Elefante Mexicana. This value is close to that observed in this study for the same forage palm cultivar.

Lima et al. (2016), on the other hand, when studying the morphological and productive characteristics of Gigante forage palm irrigated with saline water (5.25 dS m^{-1}) and subjected to cutting intensities, obtained an average cladode length of 37.87 cm for the second cultivation cycle. No significant differences were observed for this variable. The authors only observed a significant effect for average cladode width (20.95 cm), thickness (18.62 mm) and cladode area (583.46 cm^2). These values were relatively close to those of this study, however, the authors evaluated the Giant forage palm. Lima et al. (2016) did not specify whether these variables were analysed separately for primary and secondary cladodes or whether they corresponded to the average between primary and secondary cladodes. These authors concluded that the differences observed for cladode width and thickness possibly resulted from the greater efficiency of the plant's physiological and biochemical processes such as photosynthesis, respiration and transpiration, which were influenced by management practices.

Pereira et al. (2015b) evaluated the growth of forage palm clones (Orelha de Elefante Mexicana, IP A Sertânia and Miúda), in the municipality of Serra Talhada in Pernambuco, under drip irrigation with a fixed blade (7.50 mm) and three water application intervals (7, 14 and 28 days) and found that irrigation promoted better biometric increases in the clones evaluated. For Mexican Elephant Ear, the following average values were observed for first-order cladodes, in absolute terms: 23.80 cm for cladode length and 11.80 mm for cladode thickness. In relative terms, for the frequencies of 7, 14 and 28 days, these values were: 18.70 cm (7 days), 15.40 cm (14 days) and 16.20 cm (28 days) for cladode width; 68.60 cm (7 days), 59.00 cm (28 days) for cladode perimeter and; 285.00 cm^2 (7 days), 310.00 cm2 (14 days) and 377.00 cm2 (28 days). The total amount of water received by the crop was 558.00 mm (7 days),

475.00 mm (14 days) and 438.00 mm (28 days). Even with the different irrigation frequencies observed in the current study, these values were close to those of Pereira et al. (2015b) and, like these authors, there were no significant effects on the growth of primary cladodes.

Rocha et al. (2017) when studying three genotypes of irrigated fodder palm (Miúda, Orelha de Elefante Mexicana and IPA 20) and different cutting intervals for a 16-month cycle, obtained for Orelha de Elefante length, width and cladode area values of, respectively: 25.66 cm; 21.36 cm and 383.55 cm2. The results obtained in this study were higher than those found by these authors. This difference may have occurred due to the separation of the data evaluated for first and second order cladodes, while the authors possibly used the average between primary and secondary cladodes. However, this information was not indicated by Rocha et al. (2017).

Sarmento et al. (2019) in a study to assess the influence of different irrigation frequencies on the growth and productivity of Mexican Elephant Ear palm *(Opuntia stricta (*Haworth) Haworth) subjected to different irrigation frequencies (0, 7, 14 and 21 days) found an increasing linear effect on the length and width of primary cladodes as irrigation frequencies increased. This did not occur for primary cladode thickness. The averages observed for primary cladode length were: 31.16 cm (no irrigation); 32.83 cm (21 days); 34.97 cm (14 days) and; 36.07 (7 days). The average width values were: 23.86 (without irrigation); 26.46 cm (21 days); 26.73 (14 days) and; 28.90 (7 days). The thickness observed was: 19.14 mm (0 days); 17.60 mm (21 days); 18.36 mm (14 days) and; 17.32 mm (7 days). The average perimeters were 92.88 mm, 91.80 mm, 91.15 mm and 92.66 mm for 0, 21, 14 and 7 days, respectively. These results corroborate those found in this study, where the growth of the first-order cladodes was not affected, even though the different irrigation rates resulted in different frequencies, as discussed in the agrometeorological results (Table 6).

5.2.2 Secondary cladode

The summary of the analysis of variance using the F test at the 1 and 5 per cent probability level for the variables length, width, area, perimeter and thickness of secondary cladodes at 334 days after the treatments were applied is shown in Table 10.

Table 10. Summaries of the analysis of variance for secondary cladode length (SCL), secondary cladode width (SCW), secondary cladode area (SCA), secondary cladode perimeter (SCP) and secondary cladode thickness (SCTh) of Mexican Elephant Ear forage palm at 442 days of the cycle.

Source of variation	GL	Mean squares (QM)				
		CCS	LCS	ACS	PCS	ECS[1]
Blade	3	3,68^{ns}	2,09^{ns}	2573,55^{ns}	15,42^{ns}	4,63^{ns}
Salinity	3	7,18^{ns}	2,79^{ns}	4025,10^{ns}	36,87^{ns}	5,53*
Blade x Salinity	9	4,00^{ns}	4,11^{ns}	3657,90^{ns}	34,93^{ns}	1,85^{ns}
Block	3	7,11^{ns}	4,02^{ns}	5040,10^{ns}	37,58^{ns}	1,22^{ns}
Error	45	2,39	2,04	1946,91	21,93	1,76
CV (%)	-	7,26	7,65	15,32	8,02	6,53
Overall average	-	21,32	18,66	287,93	58,38	9,39

* - significant ($p<0.05$) and^{ns} - not significant ($p>0.05$) by F test^{-1} ; data transformed by square root.

According to the results of the analysis of variance using the F test (Table 10), as with the first-order cladodes, there were no effects of the treatments ($p>0.05$) on the assessed variables of length, width, area and perimeter of the secondary cladodes.

The second-order cladode variables showed average values, regardless of the treatment applied, of: 21.32 cm for cladode length, 18.66 cm for cladode width, 287.93 cm^2 for cladode area and 58.38 cm for cladode perimeter. These results were lower than those for first-order cladodes due to the morphometric characteristics of the plant itself.

Sarmento et al. (2019) obtained secondary cladode lengths for the different irrigation frequencies of: 27.77 cm (no irrigation); 27.28 cm (21 days); 28.23 cm (14 days) and; 29.26 cm (7 days). With regard to cladode width, in order from the lowest irrigation frequencies to the highest, the average values were: 21.84 cm; 24.33 cm; 24.69 cm and; 25.52 cm. The perimeters were: 87.73 cm, 83.24 cm, 85.38 cm and 86.96 cm from the lowest to the highest frequency. Despite the lack of significant effects, more frequent irrigation contributed to an increase in the growth of Mexican Elephant Ear palm. The results observed by the authors for the genus *Opuntia stricta* (Haworth) Haworth were superior to those of this study. The variables evaluated in the current study may have been influenced by the salinity of the irrigation water, which may have reduced plant growth compared to growing conditions with non-saline water, even though there was no statistical significance between the treatments.

When studying the growth of three forage palm cultivars (Orelha de Elefante Mexicana, Miúda and Baiana) subjected to nitrogen fertilisation via fertigation, Borges et al. (2015) observed average cladode lengths ranging from 25.40 to 27.52 cm in relation to the standard error for Orelha de Elefante Mexicana. It is inferred that this difference between the length obtained in this study and that observed by the authors is due to the use of average values between primary and secondary cladodes.

Donato et al. (2014) in studies on the morphometry and yield of Giant fodder palm *(Opuntia ficus-indica* Mill) under different spacings and doses of organic fertiliser found an average cladode width of 16.30 cm at 600 days after planting. This value is close to the one obtained in this study for the Orelha de Elefante cultivar, which although it is not the same cultivar studied by the authors, belongs to the *Opuntia* genus.

Sales et al. (2013) when assessing the vegetative growth of Giant forage palm at different planting densities in Curimataú Paraibano found an average value between planting densities for cladode width equal to 18.98 cm and cladode length equal to 33.89 cm, at 710 days after planting. The variable cladode area, however, was affected by the treatments with values of 440.12 cm^2 , 397.95 cm^2 and 383.05 cm^2 . The authors did not carry out any irrigation events throughout the experimental period; the water applied was only precipitation. As in the present study, the authors found no statistically significant differences between the treatments in the growth variables analysed, with the exception of plant height.

Pereira et al. (2015b) found average values for second-order cladodes equal to: 10.40 cm for length, 12.60 cm for width, 27.30 cm for perimeter and 74.60 cm2 for area. The results for these variables were lower than in this study. Statistically, in this study, the results were not significant ($p>0.05$) between the treatments applied. However, it is possible that the salinity of the irrigation water had an influence on the plant's growth, as salinity may have contributed to increasing the actual evapotranspiration (ET_r) of the forage palm due to the utilisation of sodium (Na^+) and stomatal adjustment which, in turn, tended to contribute to the plant's development. Furthermore, according to Campos (2018), the supply of water through irrigation increases the actual evapotranspiration of fodder palm compared to rainfed cultivation. In this way, the plant increases its transpiration due to the greater availability of water.

This hypothesis can be confirmed by Fonseca et al. (2019) who concluded that higher averages of forage palm morphometric characteristics under conditions of water availability indicate that even with the use of saline water, irrigation provides better conditions for crop development, perhaps by increasing photosynthetic rates.

The results showed that there were statistically significant differences ($p<0.05$) only for the thickness of the secondary cladode in relation to the salinity levels of the irrigation water. There were no effects of the blades or the interaction between the factors (blade x salinity) on the thickness of the Mexican Elephant Ear forage palm. The data showed low dispersion with a coefficient of variation (CV) of 6.53%. The other variables showed no significant effect ($p>0.05$) for any of the factors evaluated.

It was found that there was an increase in cladode thickness up to the maximum value of 9.79 mm for salinity of 1.22 dS m^{-1} , verified by the graphical adjustment equation, from which point the thickness decreased. The R^2 value from the graphical adjustment was 0.70

(Figure 13).

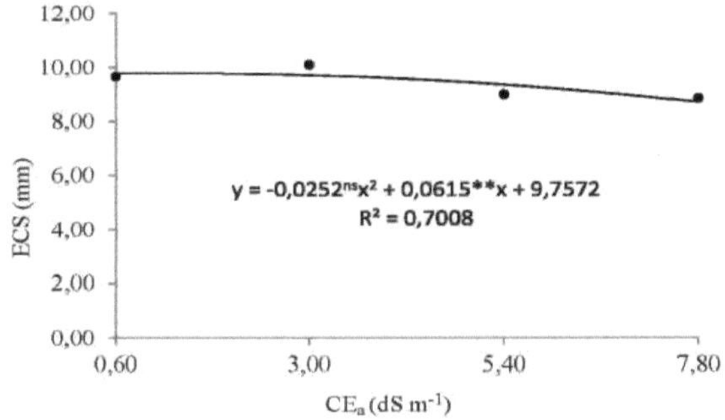

Figure 13. Secondary cladode thickness (SCS) of Mexican Elephant Ear forage palm as a function of irrigation water salinity (ECa) at 442 days of the cycle.

The thickness of the cladodia is related to the accumulation of water. According to Nobel (1991), this stored water can favour the plant's gas exchange. It is possible that this water retained in the cladodes is an indication of tolerance to the stress caused by water salinity.

The effect of the treatments ($p<0.05$) on the thickness of the secondary cladodia may indicate the development of salinity tolerance mechanisms through succulence. According to Willadino & Camara (2010), sodium tries to be transported via the xylem and accumulated in the aerial part. Therefore, it is inferred that the plant may have developed succulence to attenuate the effect of the salts accumulated in the secondary cladodes or even due to osmotic adjustment, which is responsible for causing a difference in potential and inducing the movement of water into the interior of the guard cells.

Despite this, Freire (2012) found that higher levels of salinity from 3.60 dS m^{-1} resulted in a reduction in the juiciness of forage palm. This maximum salinity level for thickness was around 60 per cent lower than that found by the author.

By associating the results obtained for cladode thickness with the agrometeorological variables, it was found that higher salt levels resulted in the maximisation of real evapotranspiration. Osmotic adjustment possibly contributed to there being no restrictions on the plant's stomata opening, which led to an increase in the transpiration component and a reduction in the cell's water content and, consequently, its thickness. This would explain the reduction in the thickness of the secondary cladode as the salinity of the irrigation water increased.

The thickness of the cladodes is one of the characteristics of the species that correlates directly with the turgidity of the plants. Thus, the greater the thickness, the greater the amount of water present in its cells, which is one of the main attributes of CAM plants (TAIZ & ZEIGER, 2013; PEREIRA et al., 2015b).

Pereira et al. (2015b) found average cladode thickness values for the cultivars OEM, IPA and Miúda of 11.80, 18.50 and 14.10 mm, respectively. The thickness observed by the authors for Mexican Elephant Ear was around 17 per cent higher.

Sales et al. (2013) obtained a cladode thickness of 23.80 mm for the Gigante forage palm, the same genus as the Mexican Elephant Ear.

Sarmento et al. (2019) found a significant effect for secondary cladode thickness as a function of irrigation frequency, which showed a decreasing quadratic effect with values of: 12.31 (0 days); 13.58 (21 days); 11.44 (14 days); 11.30 (7 days). The same did not occur for primary cladode thickness.

When comparing the results of secondary cladode thickness with Pereira et al. (2015b), Sales et al. (2013) and Sarmento et al. (2019), it was found that the thickness was lower than those found by the other authors. However, it should be borne in mind that these studies used the species *Opuntia ficus-indica* Mill, whereas this study used the species *Opuntia stricta* (Haworth) Haworth.

According to Rocha et al. (2017), cladode thickness is of great importance for the plant's photosynthetic and water storage capacity.

The Orelha de Elefante Mexicana clone has greater potential to adapt to conditions of low water availability in the soil, as it has greater water reserve capacity in its cladodes (PEREIRA et al., 2015b). This water storage capacity also makes the plant tolerant to salinity in irrigation water.

Nobel (1991) stated that mature cladodes are generally thicker, most of which is made up of whitish storage tissue.

5.2.2 Plant height and width

The summary of the results of the analysis of variance using the F test at the 1 and 5 per cent probability level for plant width and plant height as a function of the treatments with irrigation rates and irrigation water salinity are shown in Table 11.

Table 11. Summaries of the analysis of variance for plant width (PL) and plant height (PH) of Mexican Elephant Ear forage palm at 442 days of the cycle.

Source of variation	GL	Mean squares (QM)	
		LP	AP[1]
Blade	3	155,33^{ns}	51,77^{ns}
Salinity	3	28,7^{ns}	34,35^{ns}
Blade x Salinity	9	187,23^{ns}	28,84^{ns}
Block	3	159,87^{ns}	212,52^{ns}
Error	45	195,93	75,12
CV (%)	-	16,19	8,02
Overall average	-	86,44	56,03

^{ns} - not significant (p>0.05) by F test;-[1] square root transformed data.

The results showed that the treatments did not provide statistically significant differences (p>0.05) in the plant growth variables of width and height. Plant width (PL) averaged 86.44 cm and plant height (PH) averaged 56.03 cm.

Sarmento et al. (2019) also obtained similar results, in which the authors found no significant effect for the height and width of Mexican Elephant Ear palm as a function of irrigation frequencies (0, 21, 14 and 7 days). The observed values were 78.47 cm (0 days), 79.30 cm (21 days), 77.98 (14 days) and 84.67 cm (7 days) for plant height and for plant width the values were 119.86 cm (0 days), 119.07 cm (21 days), 121.21 cm (14 days) and 119.55 cm (7 days). The average plant width was close to that of this study, but the authors obtained double the value for plant height. The greater height was the result of cladodes up to the fourth order, as observed by the authors. In this study, cladodes were issued up to the second order, but the experiment was carried out in pots.

Pereira et al. (2015b) observed an average plant height of 48.80 cm after 12 months of cultivation, which was equivalent to this study.

Borges et al. (2015) obtained an average plant height of 65.00 cm for Mexican Elephant Ear.

Félix et al. (2018), in a study carried out in pots to assess the growth and production of Mexican Elephant Ear forage palm subjected to irrigation with saline water (1.50; 3.00; 4.50; 6.00 dS m^{-1}), concluded that plant height was negatively affected by salinity. The greatest heights were observed at a salinity of 1.50 dS m^{-1} . The same occurred for plant width.

Ramos et al. (2015) stated that forage palm plant height is an important plant characteristic and that it is usually related to production.

A plant width of 98.80 cm was also observed by Rocha et al. (2017) for *Opuntia stricta* (Haworth) Haworth. The authors carried out their research in field conditions under fertigation.

According to Fonseca et al. (2019), changes in crop management through irrigation indicate that, even with the use of saline water, irrigation favours an increase in the morphometric characteristics of fodder palm.

5.3 Physiology

The results of the analysis of variance using the F test at 1 and 5% probability levels for the physiological variables relative water content and electrolyte leakage are shown in Table 12.

Table 12. Summaries of the analysis of variance for relative water content (RWT) and electrical leakage (ES) of Mexican Elephant Ear forage palm at 442 days of the cycle.

Source of variation	GL	Mean squares (QM)	
		TRA	EE
Blade	3	4,47^{ns}	85,59^{ns}
Salinity	3	306,75**	33,31^{ns}
Blade x Salinity	9	52,36**	116,34**
Block	3	0,96^{ns}	75,77^{ns}
Error	45	4,92	24,53
CV (%)	-	2,44	12,87
Overall average	-	91,08	38,50

** - significant (p<0.01) and^{ns} - not significant (p>0.05) by the F test.

There were statistically significant differences (p<0.01) for relative water content (RWT) in relation to the salinity of the irrigation water and the interaction between the factors irrigation blades and saline levels. There was no significant effect (p>0.05) for the blade factor alone. For electrolyte leakage (EE) there was significance (p<0.01) only for the interaction between the factors.

5.3.1 Turgidity

According to the results obtained, the interaction between the factors for relative water content in the plant (%) had a decreasing linear effect as the salinity of the irrigation water increased for the L1 blade, and a decreasing quadratic effect for the other blades (Figure 14).

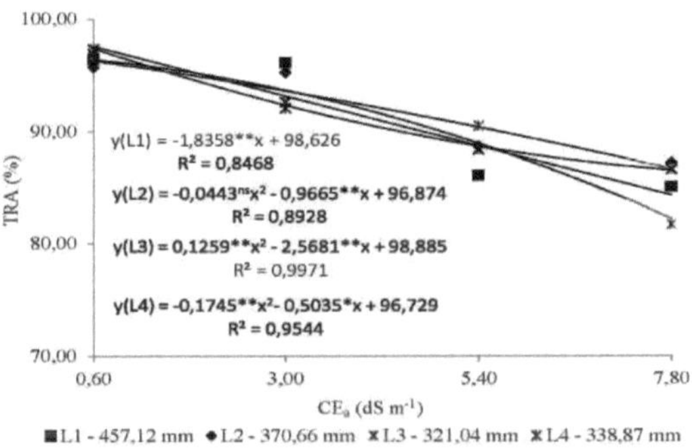

Figure 14. Relative water content (RWT) of Mexican Elephant Ear forage palm as a function of water salinity (ECa) for each irrigation rate (L1 = 25, L2 = 50, L3 = 75, and L4 = 100% of total soil water capacity - TWC) at 442 days of the cycle.

The interaction between the factors blades and salinity levels of the irrigation water showed that the maximum TRA value corresponded to water salinity tending towards zero. This could be observed for all the blades evaluated (25 per cent of the total water capacity in the soil (WCC) - L1, 50 per cent of the WCC - L2, 75 per cent of the WCC - L3 and 100 per cent of the WCC - L4).

When analysing the L1 blade (457.12 mm) at the different salinity levels, it was found that for salinity S1 (0.60 dS m^{-1}), the TRA was equivalent to 97.35% calculated using the graphic adjustment equation. The minimum TRA value, considering the range of saline levels used in the research, was obtained for salinity S4 (7.80 dS m^{-1}) and was equal to 84.14%. The percentage difference between the extremes was 13.21%.

The L2 blade (370.66 mm) resulted in a maximum TRA, within the range of treatments with salt levels, of 96.28% for S1 (0.60 dS m^{-1}) and a minimum of 86.64% for S4 (7.80 dS m^{-1}). The percentage difference between the two was 9.64%.

For the L3 blade (321.04 mm), depending on the salinity levels, it showed a maximum relative water content similar to that of the L1 blade for the same salinity of 0.60 dS m^{-1} (S1) with TRA equal to 97.39% and a minimum of 86.51%, with a difference between the values of 10.88%.

The L4 blade (338.87 mm) provided a TRA equal to 96.36% for S1 (0.60 dS m^{-1}) and a TRA equal to 82.18% for S4 (7.80 dS m^{-1}). The greatest percentage difference between the maximum and minimum values of relative water content was found for these treatments, with a percentage of 14.18.

It was therefore found that the Mexican Elephant Ear forage palm reduced its relative water content when subjected to higher salinity levels, which may be associated with stomata regulation and increased plant transpiration. This result corroborates the growth results for secondary cladode thickness, in which the thickness was reduced as the salinity of the irrigation water increased. However, even with the decrease in relative water content, the plant maintained a high water content, which can be seen in the percentage difference between the maximum (at a salinity of 0.60 dS m^{-1}) and minimum values (at a salinity of 7.80 dS m^{-1}) within the same irrigation regime (L1 - 13.21%, L2 - 9.64%, L3 - 10.88% and L4 - 14.18%).

Scalisi et al. (2016) observed that, under drought conditions, the *Opuntia* species maintains its growth and stomatal conductance with a relative water content of the cladodes of at least 45%, which is extremely low compared to other species. In this study, irrigation, even using saline water, favoured an increase in the relative water content of the cladodes compared to rainfed cultivation. Furthermore, irrigation with water with a higher electrical conductivity (7.80 dS m^{-1}) resulted in a 47% higher water content in the plant compared to non-irrigated conditions. This demonstrated the importance of adopting irrigation events in the cultivation of Mexican Elephant Ear palm and the viability, in plant terms, of using saline water in the irrigation of this crop, given that it had no significant effect ($p>0.05$) on its growth. However, it is necessary to consider the aspects of using saline water in the soil.

Barreiro Neto et al. (2017) when evaluating physiological indices in pineapple under saline stress concluded that the increase in leaf water content with the increase in irrigation water salinity is a reflection of osmotic adjustment, with the plant accumulating ions in the cell vacuole or synthesising organic compounds in order to lower the internal water potential and guarantee water absorption, keeping the cells turgid. Although the pineapple is a crop with completely different characteristics to cacti, it is possible that the decrease in water content with the increase in irrigation water salinity in the fodder palm, as observed by Barreiro Neto et al. (2017), was due to the accumulation of osmotically active solutes in the cladodes and the increase in plant transpiration.

This osmotic effect in fodder palm allows for the nocturnal acquisition of water from transpiration and dew, as well as the temporary storage of water in the vacuoles (EPIFÂNIO, 2019).

According to Almeida et al. (2019), osmotic adjustment in xerophytic forage plants promotes the maintenance of cell turgidity at very negative water potentials, which allows the cell elongation process to be maintained, in addition to greater stomatal conductance, making it possible to maximise the photosynthetic process.

Pereira et al. (2017) observed that although forage palm has a CAM morphophysiological mechanism, when studied under water deficit, the cladodes show a reduction in thickness due to the low water content of the cladodes.

Epifânio (2019), when studying resistance to mealybug attacks, nutritional value, isotopic composition and morphological descriptors of six forage palm genotypes subjected to different soil conditions, found a quadratic effect for relative water content with a daily average of 30.13%. The Mexican Elephant Ear genotype showed greater stability in relative water content values throughout the day, demonstrating a good ability to maintain constant water status in the plant. According to the author, water content is responsible for tissue turgidity and can vary depending on growing conditions. In addition, the water content stored in the cladodes of the forage palm is an important characteristic for the semi-arid region, in meeting a large part of the animals' water needs, especially during the dry period of the year, since it is already used to reduce the water intake of the animals that consume it.

5.3.2 Electrolyte leakage

Based on the results of the analysis of variance using the F test at the probability level of 1 and 5 per cent, there were significant statistical differences ($p<0.01$) for electrolyte leakage (EE) in relation to the interaction between irrigation rates and irrigation water salinity (Table 12).

The interaction between the factors showed that the maximum values of electrolyte leakage (%) were similar for the different blades, within the ranges of irrigation water salinity considered (Figure 15).

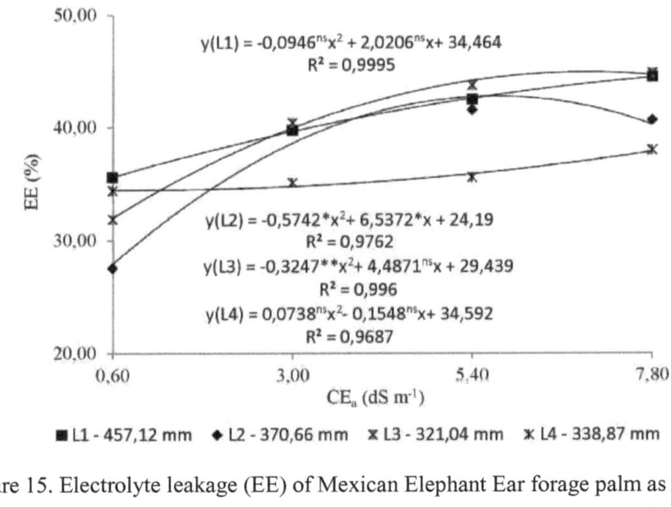

Figure 15. Electrolyte leakage (EE) of Mexican Elephant Ear forage palm as a function of water salinity (ECa) for each irrigation rate (L1 = 25, L2 = 50, L3 = 75, and L4 = 100% of total soil water capacity - CTA) at 442 days of the cycle.

It was observed that for slope L1 (25% of the total water capacity in the soil - CTA), with an accumulated value of 457.12 mm (average between treatments with slope L1), the increase in salinity of the irrigation water led to an increase in electrolyte leakage (EE,%). At this level, there was an increasing linear graphical trend for the different salinity levels with an adjustment coefficient R^2 equal to 0.9995. For the salinity of 7.80 dS m^{-1} , the graphical adjustment equation found EE = 44.47%.

An approximate value of electrolyte leakage was found for the L2 (50% of CTA) slope, with 42.79%. However, this blade showed a quadratic graphical trend as a function of the salinity levels of the irrigation water, with the maximum value observed at a salinity of 5.69 dS m^{-1} . The adjustment coefficient showed a high correlation between the data evaluated with an R^2 of 0.9762.

Similar to what was observed for L1, L3 (75% of CTA) as a function of the different electrical conductivities of the irrigation water resulted in maximum electrolyte leakage of 44.94% for the maximum salinity of 6.91 dS m^{-1} . The graphical adjustment showed decreasing quadratic behaviour and an R coefficient^2 of 0.996.

The interaction between the L4 blade (100% of CTA) and the salinity levels resulted in an increasing quadratic effect with maximum electrolyte leakage equal to 37.87% for the salinity level of 7.80 dS m^{-1} .

The results allow us to infer, therefore, that the increase in salinity of the irrigation water, regardless of the level, led to a breakdown in the integrity and increased permeability of the plasma membranes, which, according to Brito et al. (2015), is associated with a higher value of electrolyte leakage.

No research results were found in the literature directly addressing the effects of salt stress on electrolyte leakage in fodder palm plants, especially *Opuntia stricta* (Haworth) Haworth, and it was not possible to compare the values obtained with those of other authors.

Although no references were found for fodder palm, Nunes et al. (2019) observed results that corroborate those obtained. When the authors evaluated the cultivation of cowpeas (BRS Marataoã and Setentão) under different concentrations of ascorbic acid and salt levels (0.0 - control; 1.50; 3.00; 4.50; 6.00 and 7.50 dS m^{-1}), they found that treatments with saline solution caused increased membrane damage, based on electrolyte leakage. The same was observed in the present study in which increases in salt levels led to higher percentages of electrolyte leakage.

Sousa et al. (2017) in a study of scion/scion holder combinations, irrigation water salinity levels (0.60; 1.20; 1.80; 2.40 and 3.00 dS m^{-1}) and nitrogen fertilisation found increases in electrolyte leakage as salinity increased, which confirms the results observed in this study. The authors stated that this variable indicates the resistance of the cell membrane to damage that can be caused by stresses, such as salinity; therefore, the lower the values obtained for leakage, the more stable the cell membrane. The increase in the leakage of electrolytes from the cell content is due to the greater fluidity of the membrane as a result of injuries such as salt stress, a fact also observed in this study.

Ferraz et al. (2015) in an experiment with BRS Energia papaya submitted to silicon levels (0, 100, 200 and 300 mg L^{-1}) and water salinity levels (0, 2, 4 and 6 dS m^{-1}) found that increases in salinity levels caused a significant increase ($p<0.01$) in cell membrane rupture, observed by the increase in the percentage of extravasated electrolytes. The authors stated that the observed increase in electrolyte extrusion may be associated with the effect of salt phytotoxicity on plant organisms, due to the accumulation of ions in plant tissues. In addition, the disturbances caused by this accumulation consist of changes in the composition of membrane structures and cell organelles, which cause morphophysiological changes in plants. These alterations can cause an increase in the rupture of the cell membrane and consequent loss of its internal content.

Martins et al. (2018) also found that high electrolyte leakage values are related to membrane damage caused by plant stress conditions. And that membrane damage at the leaf level can be one of the first signs of stress.

5.4 Production

5.4.1 Number of primary and secondary cladodes

The results of the analysis of variance using the F test for significance at 1 and 5 per cent probability showed that there was a significant effect ($p<0.05$) only for the number of primary cladodes in relation to the irrigation rates. For the other variables of number of secondary cladodes and total number of cladodes, no statistically significant differences were observed (Table 13).

Table 13. Summaries of the analysis of variance for number of primary cladodes (NCP), number of secondary cladodes (NCS) and total number of cladodes (NTC) of Mexican Elephant Ear forage palm at 442 days of the cycle.

Source of variation	GL	Mean squares (QM)		
		NCP[1]	NCS	NCT
Blade	3	4,89*	$3{,}43^{ns}$	$1{,}64^{ns}$
Salinity	3	$2{,}89^{ns}$	$12{,}56^{ns}$	$9{,}18^{ns}$
Blade x Salinity	9	$1{,}78^{ns}$	$14{,}25^{ns}$	$14{,}52^{ns}$
Block	3	$1{,}02^{ns}$	$20{,}31^{ns}$	$15{,}10^{ns}$
Error	45	1,73	19,35	16,66
CV (%)	-	11,79	21,19	12,06
Overall average	-	5,66	11,48	17,20

* - significant ($p<0.05$) and^{ns} - not significant by F test;-[1] square root transformed data.

The number of primary cladodes showed a quadratic trend with R^2 equal to 0.8643 (Figure 16). The highest average number of primary cladodes was found, based on the graphic adjustment equation, for the 376.00 mm blade with 4.03 cladodes. This blade is the closest to L2 (370.66 mm) corresponding to 50 per cent of the total water capacity in the soil - CTA. The other blades had average values of 3.43 cladodes (L3 - 75% of the CTA equal to 321.04 mm),

3.76 cladodes (L4 - 100 per cent of CTA equal to 338.87 mm) and 2.72 cladodes (L1 - 25 per cent of CTA equal to 457.12 mm), also obtained by the fitting equation.

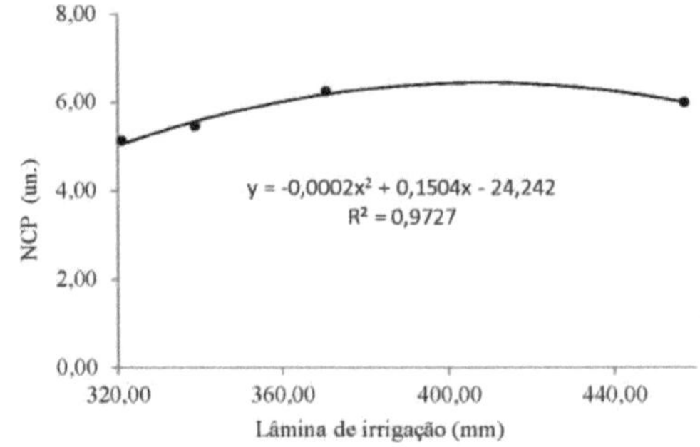

Figure 16. Number of primary cladodes (NCP) of Mexican Elephant Ear forage palm as a function of irrigation rates at 442 days into the cycle.

It was therefore found that the number of primary cladodes increased with the increase in irrigation up to 376.00 mm and then decreased.

The other variables showed no significant effect ($p>0.05$) with average values of 11.48 secondary cladodes (NCS) and 17.20 total cladodes (NTC).

Cavalcante et al. (2017) when evaluating the cultivation of Mexican Elephant Ear forage palm under the presence and absence of cladode curing found no statistical differences ($p>0.05$) for the number of primary cladodes, number of secondary cladodes and total number of cladodes. Even so, the average values were 2.35 first-order cladodes, 5.37 second-order cladodes and 12.31 total cladodes. However, the total number of cladodes was counted by the authors with the third order cladodes and in the case of this study, no tertiary cladode sprouts were observed. The values found in this study were higher than those obtained by Cavalcante et al. (2017), possibly as a result of the application of irrigation rates that led to an increase in forage palm production.

Lima et al. (2016) found no significant interaction ($p>0.05$) for the number of primary cladodes, but observed a significant effect for the total number of cladodes with a maximum value of 20.60 cladodes per plant. The authors also used saline water to irrigate fodder palm, but they assessed the productive characteristics of the Giant palm. Lima et al. (2016) concluded that a greater number of cladodes tends to increase CO_2 capture, which leads to a higher photosynthetic rate and thus helps to maximise production. This result corroborates those found in this research for Mexican Elephant Ear palm.

Sarmento et al. (2019) found cladode numbers that corroborate this research. The authors found cladode number values for Mexican Elephant Ear under different irrigation frequencies (0, 7, 14 and 21 days) of: 6.06 cladodes (0 days), 4.92 cladodes (21 days), 4.85 cladodes (14 days) and 4.82 cladodes (7 days), for first order cladodes; 12.40 cladodes (0 days), 11.91 cladodes (21 days), 11.91 cladodes (14 days) and 10.50 cladodes (7 days), for second order cladodes and; 18.60 cladodes (0 days), 17.20 cladodes (21 days), 16.94 cladodes (14 days) and 15.44 cladodes (7 days), for total number of cladodes.

According to Borges et al. (2015), a higher number of cladodes on fodder palm plants results in higher production. The authors found 13.00 cladodes per plant for Mexican Elephant Ear under fertigation, which is equivalent to the total number of cladodes.

Pereira et al. (2015b) found that water availability conditions did not affect the number of primary cladodes and the total number of cladodes of the forage palm *Opuntia stricta* (Haworth) Haworth. However, they found a significant effect on the number of secondary cladodes. The authors observed average values of: 8.11 first-order cladodes; 13.50 total cladodes per plant. The number of secondary cladodes showed values in the order of: 7.67 units (for the 7-day irrigation frequency with a total irrigated blade of 558.00 mm); 1.56 units (for the 14-day irrigation frequency with a total irrigated blade of 475.00 mm) and;4.22 units (for the 28-day irrigation frequency with a total irrigated blade of 438.00 mm). This study found an average number of secondary cladodes equal to 11.48 units for irrigation rates ranging from 321.04 mm to 457.12 mm. When comparing this result with those obtained by Pereira et al. (2015b), it can be seen that the plant made better use of the irrigation water in the production of cladodes, given that the palm produced around 33% more cladodes for approximately lower water levels than those applied by the authors.

5.4.2 Fresh mass of primary and secondary cladodes

According to the results obtained from the analysis of variance using the F-test at the 1 and 5 per cent probability levels (Table 14), significant effects were observed for the primary cladode fresh mass and secondary cladode fresh mass variables in correspondence to the isolated factors and the interaction between the factors.

Table 14. Summaries of the analysis of variance for primary cladode fresh mass (PCFM) and secondary cladode fresh mass (SCFM) of Mexican Elephant Ear forage palm at 442 days of the cycle.

Source of variation	GL	Mean squares (QM)	
		MFCP	MFCS
Blade	3	5,35**	1,02*
Salinity	3	2,94**	2,76**
Blade x Salinity	9	0,99*	1,11**
Block	3	0,13^{ns}	0,03^{ns}
Error	45	0,44	0,35
CV (%)	-	12,06	16,99
Overall average	-	5,52	3,47

**; * - significant ($p<0.01$); significant ($p<0.05$) and;^{ns} - not significant, by the F test.

Based on the analysis of variance (Table 14), the primary cladode fresh mass production variable showed significance ($p<0.01$) for the isolated factors of irrigation rates and irrigation water salinity levels. The interaction between the factors was statistically significant at a 5% probability level.

The fresh mass of the primary cladode was modelled as a function of the irrigation water levels and salinity using the response surface, which resulted in a high predictive capacity with an R value^2 of 0.92 (Figure 17).

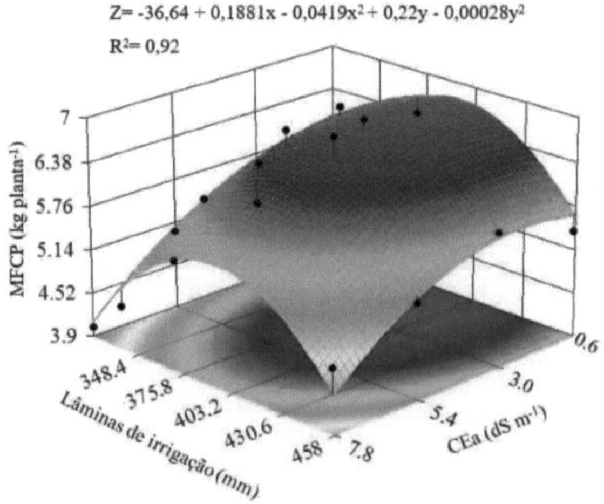

Figure 17. Fresh mass of primary cladode (MFCP) of Mexican Elephant Ear forage palm as a function of irrigation rates and electrical conductivity of irrigation water (ECa) at 442 days of the cycle.

The results indicated, according to the determination of the threshold salinity based on the equation fitting the response surface model, that the maximum fresh mass of the primary cladode (MFCP) was obtained for the salinity level of 2.20 dS m^{-1} at an irrigation depth of 392.00 mm, resulting in a production of 6.79 kg of fresh mass of the primary cladode per plant. The lowest MFCP value (4.05 kg per plant) was found for the salinity of 7.80 dS m^{-1} combined with an irrigation depth of 321.12 mm. In relation to the maximum and minimum fresh mass values, the increase of approximately 72 per cent in water salinity and 18 per cent in the irrigation depth resulted in a 40.35 per cent reduction in the fresh mass production of the primary cladode.

The same was observed by Fonseca et al. (2019) who found that the use of saline water with an electrical conductivity of 3.60 dS m^{-1} in the irrigation of Giant fodder palm was not a limiting factor for the development of the crop, probably due to the increase in photosynthetic rates. The authors stated that saline water at this electrical conductivity does not cause stress to the plant. In the present study, the salinity limit for production not to be affected was 2.20 dS m^{-1} , which is lower than the value observed by the authors, but for the Mexican Elephant Ear forage palm.

According to Barreiro Neto et al. (2017), plants that show a certain tolerance to certain abiotic stresses produce organic solutes that regulate the osmotic gradient in the cytoplasm or vacuole and are capable of producing acceptable yields under salinity conditions.

Silva (2017) when evaluating forage palm irrigated with saline water at five salinity levels (0.58 dS m^{-1} - control; 1.67 dS m^{-1} ; 2.69 dS m^{-1} ; 3.77 dS m^{-1} and; 4.78 dS m^{-1}) and three irrigation frequencies (one, two and three irrigations per week of 500 mL per plant) concluded that using water with an electrical conductivity of 4.78 dS m^{-1} is a viable option for growing the *Opuntia* and *Nopalea* genera during periods of drought. The authors also found that increasing the frequency of irrigation with salinised water from once a week to three times a week increases productivity, but increases electrical conductivity and, consequently, the concentration of salts in the soil, which could make its use and the useful life of the palm unviable.

Santos et al. (2017) obtained significant yield responses from forage palm irrigated with saline water of 3.60 dS m^{-1} , at 380 days after planting, with an increase of 111.22% in green mass production, from the rainfed condition to the highest value obtained with water application.

Rego et al. (2014) when assessing the effects of organic fertiliser and cutting intensities on the morphological characteristics and production of the Miúda palm *(Nopalea cochenillifera)* thickened and irrigated with water electrical conductivity equal to 5.25 dS m^{-1} , found an average value of 3.24 kg for primary cladode massaverde. Although the authors studied a different genus to the one used in this research, it was observed that even under conditions of salinity in the irrigation water, the forage palm was able to maintain its productive capacity due not only to the contribution of organic fertiliser, but also to the supply of water to the plant through irrigation. The same was observed in this study for the *Opuntia* genus irrigated with saline water.

According to Queiroz et al. (2015), Mexican Elephant Ear has high water use efficiency, which is reflected in its production compared to other forage palm genotypes.

When the results of the primary cladode variables were analysed, it was found that they did not show statistically significant differences for growth, but the treatments were significant for primary cladode physiology and production. It is inferred that the increase in the fresh mass of the primary cladode was due to the accumulation of water in the cladodes of the Mexican Elephant Ear palm, bearing in mind that this cactaceae is mostly made up of water, which was observed through the relative water content in the plant. In addition, although the relative water content in the plant decreased as a result of the increase in the saline level, possibly due to the increase in the transpiration rate indicated by actual evapotranspiration, it was observed that the adoption of irrigation events favoured a higher water content in the cells compared to non-irrigated fodder palm. Salinity also plays an important role in these results as it is involved in the osmotic adjustment mechanism and consequent maintenance of water absorption by the plant.

The fresh mass of the secondary cladodes showed a significant result (Table 14) for both the individual factors and the interaction between irrigation rates and salinity levels of the irrigation water.

The interaction between the factors ($p<0.01$) showed, according to the graphic adjustment equation, that the highest value of fresh mass of the secondary cladode (4.46 kg) occurred for the average accumulated irrigation level L2 of 370.66 mm (50% of the total water capacity in the soil - CTA) at a salinity of 2.64 dS m^{-1} . Increasing the salinity level at this same level led to a reduction in fresh mass, with a minimum value of 2.29 kg at a salinity of 7.80 dS m^{-1} (Figure 18).

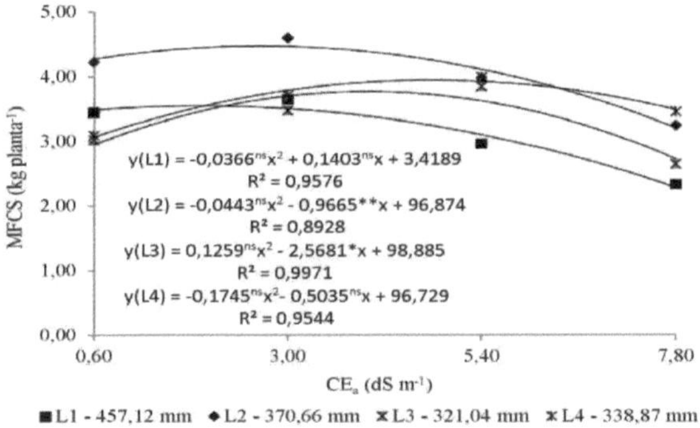

Figure 18. Fresh mass of secondary cladodia (MFCS) of Mexican Elephant Ear forage palm as a function of water salinity (ECa) for each irrigation regime (L1 = 25, L2 = 50, L3 = 75, and L4 = 100% of total soil water capacity - CTA) at 442 days of the cycle.

The other blades showed maximum secondary cladode fresh mass values of:

3.95 kg for L4 (100% of CTA) with an accumulated average of 338.87 mm and salinity of 4.77 dS m^{-1} ;

3.77 kg for L3 (75% of CTA) with a cumulative average of 321.04 mm and salinity of 3.97 dS m^{-1} and; 3.55 kg for the saline level of 1.92 dS m^{-1} , at the highest average accumulated irrigation depth L1 of 457.12 mm (25% of CTA).

The primary and secondary cladodes showed similar responses in relation to the level of salinity of the irrigation water and the irrigation depth that provided the greatest increase in fresh mass. The threshold salinity for fresh mass of the primary cladode was 2.20 dS m^{-1} at an irrigation rate of 392.00 mm and 2.64 dS m^{-1} at a rate of 370.66 mm for the second order cladode. The results obtained were consistent and the Mexican Elephant Ear palm showed a certain tolerance to irrigation water salinity.

At 480 days after transplanting, Cavalcante et al. (2017) found lower values for the green mass of primary and secondary cladodes than those observed in this study for Mexican Elephant Ear. The authors found no statistically significant differences (p>0.05) for these production variables, with average values of 0.835 kg for primary cladode green mass and 0.737 kg for secondary cladode green mass. This difference is probably due to the irrigation of the fodder palm which, in the case of the aforementioned authors, was carried out on a rainfed basis.

For the genus *Nopalea,* Rego et al. (2014) found an average fresh mass of the secondary cladode of 0.321 kg. This value is much lower than that obtained for the *Opuntia* genus in this research.

Fonseca et al. (2019) concluded that the use of saline water in forage palm irrigation did not affect its productive capacity up to the limit of 3.60 dS m^{-1} . The authors explained that this was due to the storage of ions and the accumulation of organic solutes in the plant. The authors also noted that the greater availability of water led to an increase in crop production. The same occurred in this study for the Orelha de Elefante Mexicana cultivar at salinities of up to 2.20 dS m^{-1} for the production of first-order cladodes and 2.64 dS m^{-1} for second-order cladodes. It was therefore observed that supplying the plant with water, even saline water, favoured an increase in forage production.

5.4.3 Total fresh mass and total dry mass

The results of the analysis of variance using the F test at the 1 and 5 per cent probability level for total fresh mass (MFT) and total dry mass (MST) are shown in Table 15.

Table 15. Summaries of the analysis of variance for total fresh mass (MFT) and total dry mass (MST) of Mexican Elephant Ear forage palm at 442 days into the cycle

Source of variation	GL	Mean squares (QM)	
		MFT	MST
Blade	3	3,79**	$0,04^{ns}$
Salinity	3	3,72**	$0,17^{ns}$
Blade x Salinity	9	$1,03^{ns}$	$0,12^{ns}$
Block	3	$0,13^{ns}$	$0,18^{ns}$
Error	45	0,62	0,08
CV (%)	-	8,78	22,40
Overall average	-	8,99	1,24

** - significant (p<0.01) and^{ns} - not significant, according to the F test.

The analysis of variance showed statistically significant differences ($p<0.01$) in total fresh mass (MFT) for the isolated factors of irrigation rates and irrigation water salinity levels (Table 15).

The total fresh mass showed a quadratic graphical trend for the irrigation rates, with a value of R^2 equal to 0.9133 (Figure 19).

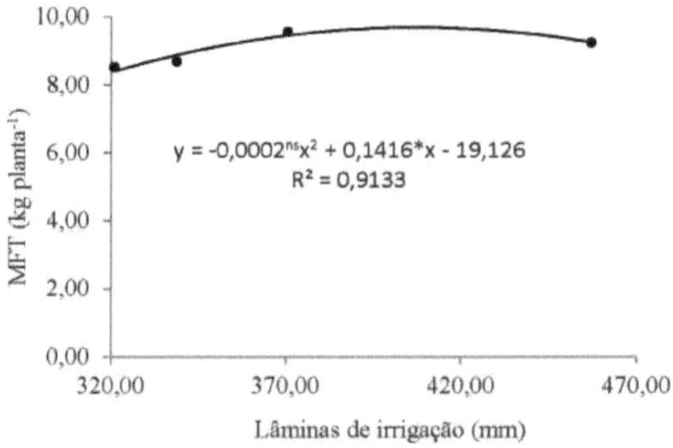

Figure 19. Total fresh mass (FFM) of Mexican Elephant Ear forage palm as a function of irrigation rates at 442 days into the cycle.

The maximum value for total fresh mass, determined using the graphic adjustment equation, as a function of the irrigation rate was found for the 354.00 mm rate, equal to 5.94 kg. Increasing the total irrigation depth above 354.00 mm led to a decrease in fresh mass.

The lowest total fresh mass value was observed with an irrigation depth of 457.12 mm, equivalent to treatment L1 (25% of total soil water capacity), with a fresh mass of 3.81 kg.

With regard to the effect of the saline levels of the irrigation water on the total fresh mass, the threshold salinity was 2.23 dS m^{-1} resulting in a production of 9.34 kg, according to the adjustment equation. The lowest production occurred at the highest salinity level used in the interval between treatments (7.80 dS m^{-1}) and resulted in a value of 8.36 kg (Figure 20).

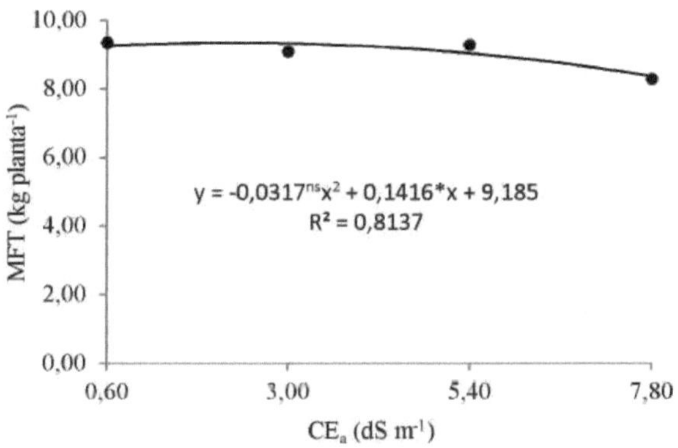

Figure 20. Total fresh mass (MFT) of Mexican Elephant Ear forage palm as a function of irrigation water salinity levels (ECa) at 442 days of the cycle.

It was possible to observe that even though the salinity of the irrigation water increased by 71.41 per cent, this increase resulted in a decrease of only 10.49 per cent in the production value, showing that it is feasible to use water with salinity above the threshold of up to 7.80 dS m^{-1} .

On many farms in the northeastern semi-arid region, producers only have water with high electrical conductivity available, and it cannot be used for other agricultural crops. It would therefore be feasible to use this water to irrigate fodder palm, but the soil's physical and chemical characteristics must be taken into account so as not to render it unusable in the future.

Under the conditions of this study, the interaction between the factors for total fresh mass had no significant effect ($p>0.05$). The treatments L1S3 (25% of the total water capacity in the soil - CTA with an electrical conductivity of 5.40 dS m^{-1}), L1S4 (25% of the CTA and 7.80 dS m^{-1}), L2S3 (50% of the CTA and 5.40 dS m'1), L2S4 (50% of the CTA and 7.80 dS m'1), L3S3 (75% of the

CTA and 5.40 dS m^{-1}), L3S4 (75% of CTA and 7.80 dS m^{-1}) in productive terms did not affect the total fresh mass of the Mexican Elephant Ear forage palm, but they did have negative consequences for the soil. According to the final soil analysis (Table 4 and Figure 4), these treatments resulted in saline sodic soils. Therefore, for the use of water with high salt contents combined with irrigation rates, in the specific case of total fresh mass and under the conditions in which the research was carried out, treatments L1S1 (25% of CTA and 0.60 dS m^{-1}), L1S2 (25% of CTA and 3.00 dS m^{-1}), L2S1 (50% of CTA and 0.60 dS m^{-1}), L2S2 (50% of CTA and 3.00 dS m- 1), L3S1 (75% CTA and 0.60 dS m^{-1}), L3S2 (75% CTA and 3.00 dS m^{-1}), L4 at salinities S1, S2 S3 and S4 (100% CTA and 0.60 dS m^{-1} , 3.00 dS m^{-1} , 5.40 dS m^{-1} and 7.80 dS m^{-1} , respectively) would be the most suitable, as for these treatments the soil samples were classified as normal (non-saline and non-sodic) at the end of the experimental period. The other treatments could also be used, but leaching fractions would have to be adopted to mitigate the effect of the accumulation of salts in the irrigation water on the soil.

The greatest contribution to the total fresh mass of cladodes came from the fresh mass of primary cladodes, as observed in the results found for primary and secondary cladodes. Pinheiro et al. (2014) corroborate this result when they state that OEM has semi-open growth that favours the emission of primary cladodes, with a large contribution from first-order cladodes in the final production.

Silva et al. (2015d) compared the productivity of IPA Sertânia, Miúda and Orelha de Elefante Mexicana forage palm cultivars resistant to carmine mealybug and observed that the Orelha de Elefante Mexicana cultivar stood out as the most productive genotype in terms of fresh mass in the semi-arid region.

The total dry mass (TDS) did not show a significant effect for any of the factors evaluated ($p>0.05$), with an average value of 1.24 kg. It is inferred that the statistical difference observed in total fresh mass was due to the storage of water in the cladodes of the Mexican

Elephant Ear palm in response to the treatments applied; however, it was found that the treatments did not influence the accumulation of dry mass in the forage. This indicated that the forage palm remained with a high water content as a result, perhaps, of the development of tolerance mechanisms to the salinity of the irrigation water and also due to the supply of water to the plant.

Rocha et al. (2017) when evaluating the effects of cutting intervals on the productive characteristics of forage palm genotypes under irrigated conditions, obtained for Mexican Elephant Ear averages of 208,88 tonnes of green mass per hectare (t MV ha^{-1}), 323.75 t MV ha^{-1} , 566.00 t MV ha^{-1} and 763.50 t MV ha^{-1} at 4, 8, 12 and 16 months of cutting, respectively. This corresponded to 4.18 kg, 6.47 kg, 11.32 kg and 15.27 kg. A total of 738.50 mm of water was supplied to the plant, including precipitation and irrigation. In the current study, the total fresh mass was determined 334 days after the treatments were applied, which corresponded to a total cycle of 442 days, with a maximum production of 9.34 kg. This figure was lower than that found by the authors for an equivalent period. However, the authors grew oil palm under field conditions. Rocha et al. (2017) also found a maximum value of 1.52 kg of dry mass per plant, cut at 12 months of cultivation and stated that this result was due to the greater volume of water applied. This value is similar to the one obtained in this study.

The results obtained for total fresh mass and total dry mass were higher than those found by Silva et al. (2015a) for cultivation under field conditions. When the authors assessed the growth and productivity of forage palm clones in the semi-arid region and their relationship with meteorological variables, they found that the Mexican Elephant Ear cultivar produced 163.70 tonnes of fresh matter per hectare and 15.60 tonnes of dry matter per hectare, with a planting density of 31250 plants per hectare. This corresponded to 5.24 kg of fresh mass per plant and 0.50 kg of dry mass per plant, respectively. It should be emphasised that the cultivation carried out by the authors was based solely on rainfall, and no irrigation events were used.

Queiroz et al. (2015) evaluated the morphophysiological characteristics and productivity of Mexican Elephant Ear palm under different irrigation rates (0.0, 8.75, 17.50, 26.35 and 35.00% of reference evapotranspiration) and found an average dry mass per plant of 0.52 kg at 380 days after planting. Even under irrigated conditions and for the same cultivar used in this study, the authors found a lower dry mass value.

Santos et al. (2017) obtained significant yield responses from forage palm irrigated with saline water of 3.60 dS m^{-1} , at 380 days after planting, with a 68.40% increase in dry mass production, from the rainfed condition to the highest value obtained with water application.

Fonseca et al. (2019) concluded that the use of irrigation with saline water applied at 33% of the reference evapotranspiration at 3-day intervals significantly increases fodder palm production and is considered viable for the sustainability of agricultural production. They also stated that the greater availability of saline water promoted an increase in crop production and, consequently, greater water use efficiency. On the other hand, conditions of low water availability are associated with lower production values, which is related to lower plant growth.

6. CONCLUSIONS

1. Applying irrigation rates of between 320.00 and 460.00 mm favours the actual evapotranspiration of the Mexican Elephant Ear palm *(Opuntia stricta (*Haworth) Haworth) for a 15-month cycle.

2. Even with irrigation water salinity of up to 7.80 dS m^{-1} , the plant maintains a satisfactory evapotranspirometric demand.

3. For the conditions of water availability and irrigation water salinity adopted in this research, the crop coefficient of Mexican Elephant Ear palm ranged from 0.68 to 0.85.

4. The growth of Mexican Elephant Ear forage palm plants was not influenced by irrigation rates and irrigation water salinity, with the exception of the thickness of second-order cladodes, which was positively affected by the electrical conductivity of the water.

5. Increasing the electrical conductivity of the irrigation water led to a reduction in the relative water content of the Mexican Elephant Ear plant.

6. High levels of electrical conductivity in irrigation water lead to an increase in the percentage of electrolytes released from the *Opuntia stricta* (Haworth) Haworth species.

7. The threshold salinity of the Mexican Elephant Ear palm *(Opuntia stricta (*Haworth) Haworth) of 2.23 dS m^{-1} obtained in this study and the irrigation depth of 354.00 mm resulted in the highest total fresh mass production values of 9.34 kg and 5.94 kg respectively.

BIBLIOGRAPHICAL REFERENCE

Adorian, G. C.; Lorenconi, R.; Dourado Neto, D.; Reichardt, K. Potential evapotranspiration and crop coefficient of two upland rice genotypes. Revista de Agricultura, v.90, p. 128-140, 2015.

Allen, R. G. Standard Reference Evapotranspiration Calculations: REF-ET. In: Irrigation and Drainage: Saving a Threatened Resource. In Search of Solutions. ASCE, 1992. p.140-145.

Allen, R. G.; Pereira, L. S.; Raes, D.; Smith, M. Crop evapotranspiration: guidelines for computing crop water requirements. Rome: Food and Agriculture Organisation, 1998. 297p, v.56.

Almeida, I. V. B.; Souza, J. T. A.; Batista, M. C. Genetic improvement of xerophilous fodder plants: Review. Publications in Veterinary Medicine and Animal Science, v.13, p.153, 2019.

Amorim, D. M.; Silva, T. G. F. D.; Pereira, P. D. C.; Souza, L. S. B. D.; Minuzzi, R. B. Phenophases and cutting time of forage cactus under irrigation and cropping systems. Pesquisa Agropecuária Tropical, v.47, p.62-71, 2017.

Anower, M. R.; Peel, M. D.; Mott, I. W.; Wu, Y. Physiological processes associated with salinity tolerance in an alfalfa half-sib family. Journal of Agronomy and Crop Science, v.203, p.506518, 2017.

Barbosa, M. L.; Silva, T. G. F. da; Zolnier, S.; Silva, S. M. S. e; Morais, J. E. F. de; Assis, M. C. de S. Association of morphological and water factors with irrigated forage cactus yield. Brazilian Journal of Agricultural and Environmental Engineering, v.21, p.600-605, 2017.

Barreiro Neto, M.; F, P. D.; Gheyi, H. R.; Lacerda, J. T.; Carvalho, R. A. Physiological indices in pineapple genotypes under salt stress. Agricultural Technology & Science, v.11, n.6, p.89-99, 2017.

Basu, S.; Giri, R. K.; Benazir, I.; Kumar, S.; Rajwanshi, R.; Dwivedi, S. K.; Kumar, G. Comprehensive physiological analyses and reactive oxygen species profiling in drought tolerant rice genotypes under salinity stress. Physiology and Molecular Biology of Plants, v.23, p.837-850, 2017.

Bernardo, S.; Soares, A. A.; Mantovani, E. C. Irrigation Manual. 8 ed. Viçosa: Ed. UFV, 2006. 625p.

Bispo, R. de C.; Hernandez, F. B. T.; Teixeira, A. H. de C. Water balance and estimation of relative water consumption of the sugarcane crop in the Northwest region of São Paulo. Irriga, v.1, p.94101, 2017.

Borges, V. E.; Matos, R. M.; Silva, P. F.; Dantas Neto, J.; Sobrino, T. G. Growth analysis of three forage palm cultivars. Higher Agricultural Education, v.30, p.7-14, 2015.

Brito, C. H.; Lopes Batista, E.; Albuquerque, I. de; Luna Batista, J. de. Evaluation of alternative products and pesticides in the control of the cochineal in Paraíba. Revista de Biologia e Ciências da Terra, v.8, p.1-5, 2008.

Brito, M. E. B.; Anjos Soares, L. A. dos; Lima, G. S. de; Silva Sá, F. V. da; Araújo, T. T.; Silva, E. C. B da. Growth and phytomass formation of tomato plants under water stress in the phenological phases. Irriga, v.20, p.139-153, 2015.

Cairo, P. A. R. Basic Course in Plant Water Relations. Universidade Estadual Do Sudoeste Da Bahia, Departamento de Fitotecnia e Zootecnia: Vitória Da Conquista, 1995, 32p.

Camargo, Â. P.; Sentelhas, P. C. Evaluation of the performance of different methods for estimating potential evapotranspiration in the State of São Paulo, Brazil. Revista Brasileira de Agrometeorologia, v.5, p.89-97, 1997.

Campos, A. R. F. Irrigation management in forage palm: definition of criteria based on the matric potential of water in the soil. Cruz das Almas: UFRB, 2018. 102p. PhD Thesis.

Campos, F. S.; Gois, G. C.; Vicente, S. L. A.; Macedo, A. de; Matias, A. G. da S. Alternative forage for goats and sheep raised in the semi-arid region. Nutritime Revista Eletrônica, v.14, p.50045013, 2017.

Carvalho, D. F. de; Oliveira, L. F. C. de. Water planning and management in irrigated agriculture. 1.ed. Viçosa: UFV, 2012. 240p.

Carvalho, L. C. de; Rios, G. F. A.; Miranda, W. L.; Castro Neto, P. Reference evapotranspiration: a current approach to different estimation methods. Pesquisa Agropecuária Tropical, v.41, p.456-465, 2011.

Cavalcante, A. B.; Leite, M. L. M. V.; Pereira, J. S.; Lucena, L. R. R. Forage palm growth as a function of curing cladode segments. Agricultural Technology & Science, v.11, p.15-20, 2017.

Clark, G. A.; Albregts, E.; Stanley, C. D. Water requirements and crop coefficients of dripirrigated strawberry plants. Transaction of ASAE, v.39, p.905-912, 1996.

Cruz Neto, J. F.; Morais, J. E. F.; Souza, C. A. A.; Carvalho, H. F. de S.; Rodrigues, C. T.

A. Silva, T. G. F. da. Applicability of agrometeorological indicators for analysing water increment by irrigation in fodder palm production systems, cv. Miúda. Journal of Environmental Analysis and Progress, v.2, p.98-106, 2017.

Divincula, J. S.; Silva, C. B. da; Santos, M. A. L. dos; Santos, D. P dos; Santos, L. W.

dos. Crop coefficient and water requirement of fodder palm in the agreste of Alagoas. Brazilian Journal of Agricultural and Environmental Engineering, v.23, p. 925-929, 2019.

Donato, P. E. R.; Pires, A. J. V.; Donato, S. L. R.; Bonomo, P.; Silva, J. A.; Aquino, A. A. Morphometry and yield of 'Gigante' forage palm under different spacing and doses of organic fertiliser. Revista Brasileira de Ciências Agrárias, v.9, p.151-158, 2014.

Epifânio, N. M. L. S. Morphological descriptors and indicator of photosynthetic metabolism in forage palm clones. Recife: UFRP, 2019. 84p. PhD Thesis.

FAO - Food and Agriculture Organisation of the United Nations. Agroecology, cultivation and uses of fodder palm. Rome: FAO, 1995. 216p.

Félix, E. S.; Lima, W. B.; Silva, C. T.; Araújo, J. S.; Pereira, D. D.; Lira, E. C. Cultivation of forage palm *(Opuntia stricta)* irrigated with salinised water. Brazilian Applied Science Review, v.2, p.1869-1875, 2018.

Ferraz, R. L. de S.; Magalhães, I. D.; Beltrão, N. E. de M.; Melo, A. S. de, Brito Neto, J. F. de; Rocha, M. do S. Photosynthetic pigments, cell extrusion and relative water content in papaya leaves under silicon and salinity. Revista brasileira de engenharia agrícola e ambiental, v.19, p.841-848, 2015.

Ferreira, D. F. SISVAR - a computer statistical analysis system. Ciência e Agrotecnologia, v.35, p.1039-1042, 2011.

Fonseca, V. A.; Santos, M. R. dos; Silva, J. A. da, Donato, S. L. R., Rodrigues, C. S.; Brito, C. F. Ba. Morpho-physiology, yield, and water-use efficiency of *Opuntia ficus-indica* irrigated with saline water. Acta Scientiarum Agronomy, v.41, p.1-11, 2019.

Francisco, P. R. M.; Medeiros, R. M. de; Matos, R. M. de; Bandeira, M. M.; Santos, D. Analysis and mapping of the humidity, water and aridity indices using the BHC for the State of Paraíba. Revista Brasileira de Geografia Física, v.8, p.1093-1108, 2015.

Franco-Salazar, V. V.; Véliz, J. A. Efectos de la salinidad sobre el crecimiento, acidez titulable y concentrtación de clorofila en *Opuntia ficus-indica* (L.) MILL. Saber, University of Oriente, v.20, p.12-17, 2008.

Freire, J. L. de. Evaluation of forage palm clones *(Opuntia* and *Nopalea)* under irrigation and salinity. Pernambuco: UFRP, 2012. 85p. PhD Thesis.

Frota, M. N. L. da; Carneiro, M. S. S de; Carvalho, G. M. C.; Araújo Neto, R. B de. Fodder palm in animal feed. Teresina: Embrapa Meio-Norte, 2015. 47p.

Gava, C. A. T.; Lopes, E. B. Production of forage palm seedlings using cladode fragments. Petrolina: Embrapa Semiárido, 2012. 2p.

Gheyi, H. R.; Dias, N. da S.; Lacerda, C. F. de. Salinity management in agriculture:

basic applied studies. Fortaleza: INCT Sal, 2010. 472p.

Jardim, A. M. da R. F.; Silva, M. J. da.; Queiroz, M. G. de; Alves, C. P.; Silva, T. G. F. da. Evapotranspiration and crop coefficient of *Opuntia stricta* (Haw.) Haw. in the fourth productive cycle. XX Brazilian Congress of Agrometeorology and V Symposium on Climate Change and Desertification in the Brazilian Semi-Arid. Juazeiro-BA/Petrolina-PE, Brazil: 14 to 18 August 2017.

Larcher, W. Carbon utilisation and dry matter production. In: Larcher, W. Ecologia vegetal. São Paulo: EPU, 1986. p.74-160.

Libardi, P.L. Dynamics of water in the soil. Piracicaba: EDUSP, 1995. 497p.

Lima, G. F. da C.; Rego, M.; Teles, M.; Dantas, F. D. G.; Lôbo, R. N. B.; Silva, J. G. M. D.; Aguiar, E. M. D. Morphological characteristics and forage productivity of irrigated cactus pear under different cutting intensities. Revista Caatinga, v.29, p.481-488, 2016.

Lima, G. F. da C.; Wanderley, A. de M.; Guedes, F. X.; Rego, M. M. T.; Dantas, F. D. G.; Silva, J. G. M da; Novaes, L. P; Aguiar, E. M de. Irrigated and densified fodder palm: a strategic fodder reserve for the Potiguar semi-arid region. Agricultural Research Company of Rio Grande do Norte - EMPARN, Parnamirim, Rio Grande do Norte, 2015. Embrapa Semiárido (Document 45)

Lima, T. P. de; Gomes, R. R. F.; Reis E. F.; Carvalho, C. M. de; Cadore, R. Freitas, D. S. Production of pepper Capsicum chinense under different irrigation depths in greenhouse. Brazilian Journal of Irrigated Agriculture, v.11, p.1254-1260, 2017.

Lopes, A. S.; Pavani, L. C.; Corá, J. E.; Zanini, J. R.; Miranda, H. A. Irrigation management (tensiometry and climatological water balance) for the bean crop in direct and conventional cultivation systems. Engenharia Agrícola, v.24, p.89-100, 2004.

Lopes, O. D.; Kobayashi, M. K.; Oliveira, F. G.; Alvarenga, I. C.; Martins, E. R.; Corsato, C. E. Determination of crop coefficient (Kc) and water use efficiency of irrigated pepper rosemary. Revista Brasileira de Engenharia Agrícola e Ambiental, v.15, n.6, p.548-553, 2011.

Matos, R. M.; Neto, J. D.; de Lima, A. S.; da Silva, P. F.; Borges, V. E.; Sobrinho, T. G. Moisture content by different methods in a neosol of the Brazilian semiarid. Brazilian Journal of Irrigated Agriculture, v.11, n.4, p.1588, 2017.

Martins, A. C.; Larré, C. F.; Bortolini, F.; Borella, J.; Eichholz, R.; Delias, D.; Amarante, L do. Water deficit tolerance: differential adaptation between forage species. Iheringia. Botanical Series, v.73, p.228-239, 2018.

Morais, J. E. de; Silva, T. G. da; Queiroz, M. G. D.; Araujo, G. G. de; Moura, M. S.; Araújo Júnior, G. D. N. Hydrodynamic changes of the soil-cactus interface, effective actual evapotranspiration and its water efficiency under irrigation. Brazilian Journal of Agricultural and Environmental Engineering, v.21, n.4, p.273-278, 2017.

Nobel, P. S. Achievable productivities of CAM plants: basis for high values compared with C3 and C4 plants. New Phytol, v.119, p.183-205, 1991.

Novais, R. F.; Neves, J. C. L.; Barros, N. F. Trials in a controlled environment. In: Oliveira, A. J. (ed.) Research methods in soil fertility. Brasília: Embrapa SEA, 1991. 189-25p.

Nunes, L. R. de L.; Pinheiro, P. R.; Cabral, F. A. S.; Silva, J. B. da; Dutra, A. S. Ascorbic acid in cowpea seeds under salt stress. Journal of Seed Science, v.41, p.441- 451, 2019.

Oliveira, F. I.; Medeiros, W. J. de; Lacerda, C. F. D.; Neves, A. L.; Oliveira, D. R. Saline water irrigation managements on growth of ornamental plants. Brazilian Journal of Agricultural and Environmental Engineering, v.21, p.739-745, 2017.

Oliveira, F. T. de; Souto, J. S.; Silva, R.P. da; Andrade Filho, F.C. de; Pereira Júnior, E. B. Forage palm: adaptation and importance for arid and semi-arid ecosystems. Revista Verde de Agroecologia e Desenvolvimento Sustentável, v.5, p.27-37, 2010.

Pedrotti, A.; Chagas, R. M.; Ramos, V. C.; Prata, A. P. do N.; Lucas, A. A. T.; Santos, P. B. dos. Causes and consequences of soil salinisation. Electronic Journal on Environmental Management, Education and Technology, v.19, p.1308-1324, 2015.

Pereira, A. R.; Villa Nova, N. A.; Sediyama, G. C. Evapotranspiration. Piracicaba: FEALQ, 1997. 183p.

Pereira, P. C.; Silva, T. G, F.; Zolnier, S.; Morais, J. E. F.; Santos, D. C. Soil water dynamics and evapotranspiration of forage cactus clones under rainfed conditions. Pesquisa Agropecuária Brasileira, Goiânia, v.50, p.515-525, 2015a.

Pereira, P. C.; Silva, T. G. F.; Zolnier, S.; Morais, J. E. F.; Santos, D. C. Morphogenesis of drip-irrigated forage palm. Revista Caatinga, v.28, n.3, p.184-195, 2015b.

Pereira, P. D. C.; Silva, T. G. F. D.; Zolnier, S.; Silva, S. M. S. E.; Silva, M. J. D. Water balance in soil cultivated with forage cactus clones under irrigation. Revista Caatinga, v.30, p.776785, 2017.

Pinheiro, K. M.; Silva, T. G. F.; Sousa Carvalho, H. F.; Santos, J. E. O.; Morais, J. E. F.; Zolnier, S.; Santos, D. C. Correlations of cladode area index with morphogenic and productive characteristics of forage palm. Pesquisa Agropecuária Brasileira, v.49, p.939-947, 2014.

Prevedello, C. L. Soil physics with solved problems. Curitiba: C.L. Prevedello, 1996. 446 p.

Queiroz, J. E.; Goçalves, A. C. A.; Souto, J. S.; Folegatti, M. V. Evaluation and monitoring of soil salinity. In Gheyi, R. R.; Dias, N. S.; Lacerda, C. F. Management of salinity in agriculture: basic and applied studies. Fortaleza: INCT Sal, 2010. 472p.

Queiroz, M.G.; Silva, T. G. F.; Zolnier, S.; Silva, S. M. S. E.; Lima, L. R.; Alves, J. O. Morphophysiological characteristics and productivity of forage palm under different irrigation rates. Brazilian Journal of Agricultural and Environmental Engineering, v.19, p.931-938, 2015.

Queiroz, M.G.; Silva, T. G. F.; Zolnier, S.; Silva, S. M. S.; Souza, C. A. A.; Carvalho, H. F. S. Water-economic relationships of forage palm grown in a semi-arid environment. Irriga, v.1, p.141-154, 2016.

Ramos, J. P. R.; Leite, M. L. M. V.; Oliveira Júnior, S.; Nascimento, J. P.; Santos, E. M. Vegetative growth of *Opuntia ficus-indica* at different planting spacings. Revista Caatinga, v.24, p.41-48, 2015.

Rego, M. M. T.; Lima, G. F. C.; Silva, J. G.M.; Guedes, F. X.; Dantas, F. D. G.; Lobo, R. N.

B. Morphology and biomass yield of irrigated small palm under organic fertiliser doses and cutting intensities. Scientific Journal of Animal Production, v.16, p.118-130, 2014.

Reichardt, K. Water in agricultural systems. São Paulo: Manole, 1990. 188p.

Resende, R. S.; Amorim, J. R. A.; Cruz, M. A. S.; Meneses, T. N. Spatial distribution and natural leaching of salts in soils of the California Irrigated Perimeter, Sergipe. Brazilian Journal of Agricultural and Environmental Engineering, v.18, p.46-52, 2014.

Ribeiro, A. A.; Moreira, F. J. C.; Seabra Filho, M.; Menezes, A. S. Emergence of yellow passion fruit under saline stress in different substrates. Brazilian Journal of Biosystems Engineering, v.10, p.27-36, 2016a.

Ribeiro, A. A.; Simeão, M.; Santos, A. R. B. Comparison of methods for estimating reference evapotranspiration in the rainy and dry seasons in Piripiri (PI). Revista Agroambiental, v.8, p. 89-100, 2016b.

Richards, L.A. Diagnosis and improvement of saline and alkali soils. Washington: US Department of Agriculture, 1954. 160p.

Rocha, J. E. da S. Palma forrageira no Nordeste do Brasil o estado da arte. 21.ed. Sobral: Embrapa Caprinos e Ovinos, 2012. 40p.

Rocha, R. S.; Voltolini, T. V.; Gava, C. A. T. Productive and structural characteristics of irrigated forage palm genotypes at different cutting intervals. Archivos de Zootecnia, v. 66, p. 363-371, 2017.

Salassier, B., Soares, A. A.; Mantovani, E. C. Manual de irrigação. 8 ed. Viçosa: Ed. UFV,

2006. 625p.

Sales, A. T.; Leite, M. L. M. V.; Alves, A. Q.; Ramos, J. P. F.; Nascimento, J. P. Vegetative growth of forage palm at different planting densities in Curimataú Paraibano. Agricultural Technology & Science, v.7, p.19-24, 2013.

Santana, M. J. de; Wachsmuth, R.; Sgobi, M. A.; Almeida, F. da S. Evapotranspiration and crop coefficient of tifton-85 in Uberaba, MG. Global Science and Technology, v.9, p 39-50,

2 016.

Santos, D. C. dos; Farias, I.; Lira, M. de A.; Santos, M. V. F. dos; Arruda, G. P. de; Coelho, R. S. B.; Dias, F. M.; Melo, J. N. de. Management and utilisation of fodder palm *(Opuntia* and *Napolea)* in Pernambuco. Recife: IPA, 2006. 48p.

Santos, M. A. A. R.; Santos, G. P. S.; Lima, C. B. da; Silva, D. P.; Santos, M. A. L. Crop coefficient (kc) of fodder palm with irrigation shift. XXV CONIRD - National Congress on Irrigation and Drainage: UFS - São Cristóvão/SE, 2015. 144-1149p.

Santos, M. R. dos; Silva, A. J. P. da; Fonseca, V. A.; Campos, A. R. F.; Lisboa, M. de A. Irrigation in forage palm. Informe Agropecuário, Belo Horizonte, v.38, n.296, p.00-00, 2017.

Sarmento, W. G. C.; Oliveira, J. F. F. de; Cardoso, D. B.; Santos, E. A. dos; Silva Mendonça,

C. G. da; Andrade, A. P. de; Melo, A. A. S. De. Growth and productivity of Mexican elephant ear palm *(Opuntia stricta Haw)* subjected to different irrigation frequencies. XXVII SEMANA DE ZOOTECNIA DA UFRPE: RECIFE, 2019.

Scalasi, A.; Morandi, B.; Inglese, P.; Lo Bianco, R. Cladode growth dynamics in *Opuntia ficus-indica* under drought. Environmental and Experimental Botany, v.122, p.158-167, 2016.

Sediyama, G. C. Water requirements for crops. Brasília: ABEAS, 1987. 143p.

Seki, K. SWRC fit - a nonlinear fitting program with a water retention curve for soilshaving unimodal and bimodal pore structure. Hydrol. Earth System Science Discussion, v.4, p.407- 437,

2 007.

Silva, A. A. R. D.; Lima, G. S. D.; Azevedo, C. A. V. D.; Gheyi, H. R.; Souza, L. D. P.; Veloso, L. L. D. S. A. Gas exchanges and growth of passion fruit seedlings under salt stress and hydrogen peroxideL Pesquisa Agropecuária Tropical, v.49, e55671, 2019.

Silva, A. O. da; Klar, A. E.; Silva, E. F. de F. e; Tanaka, A. A.; Júnior, J. F. S. Water relations in sugar beet cultivars at different levels of soil salinity. Revista Brasileira de Engenharia Agrícola e Ambiental, v.17, p.1143-1151, 2013a.

Silva, A. O.; Silva. Ê. F. F.; Klar, A. E. Water use efficiency in sugar beet cultivars subjected to different soil water tensions. Water Resources and Irrigation Management, v.2, p.27-36, 2013b.

Silva, G. F.; Miranda, K. R de.; Santos, D. C. dos.; Queiroz, M. da C.; NETO, J. F. da C.; Araújo, J. E. M. Cladode area of fodder palm clones: modelling, analysis and applicability. Revista Brasileira de Ciências Agrárias, v.9, p.633-641, 2014a.

Silva, L. M. da; Fagundes, J. L; Viegas, P. A. A.; Muniz, E. N.; Rangel, J. H. de A.; Moreira, A. L.; Backes, A. A. Productivity of forage palm grown at different planting densities. Ciência Rural, v.44, p.2064-2071, 2014b.

Silva, M. L. R. B.; Figueirôa, C. S.; Mergulhão, A. C. E. S.; LYRA, M. C. C. P. Identification of diazotrophic bacteria isolated in oil palm cultivars *(Opuntia* and *Nopalea)* using the recA gene. Bioscience Journal, v. 31, p. 577-583, 2015d.

Silva, N. F. da; Cunha, F. N.; Teixeira, M. B.; Soares, F. A. L.; Sousa, C. V. de. Temporal variation of the matric potential of a latosol monitored by tensiometry. Brazilian Journal of Irrigated Agriculture, v.1, 2017a.

Silva, P. F.; Matos, R. M.; Borges, V. E.; Melo, J. A. P.; Dantas, N. J. Morphological characteristics of three forage palm cultivars under mineral fertilisation in Campina Grande-PB. Revista Enciclopédia Biosfera, v. 11, p.385-397, 2015c.

Silva, R. H. D. Growth of forage palm irrigated with saline water. Viçosa: UFV, 2017. 66pc.

Silva, T. G. F.; Araújo Primo, J. T.; Morais, J. E. F. de; Diniz, W. J. da S.; Souza, C. A. A. de; Silva, M. da C. Growth and productivity of forage palm clones in the semi-arid region and relationships with meteorological variables. Revista Caatinga, v.28, p.10-18, 2015a.

Silva, T. G. F.; Araújo Primo, J. T.; Moura, M. S. B.; Silva, S. M. S.; Morais, J. E. F.; Pereira, P. C.; Souza, C. A. A. Soil water dynamics and evapotranspiration of forage cactus clones under rainfed conditions. Pesquisa Agropecuária Brasileira, v.50, p.515-525, 2015b.

Silva, T. G. F.; Araújo Primo, J. T.; Silva, S. M. S.; Moura, M. S. B.; Santos, D. C.; Silva, M. C; Araújo, J. E. M. Indicators of water and nutrient use efficiency of forage palm clones under rainfed conditions in the Brazilian semiarid. Revista Bragantia, v.73, p.184-191, 2014c.

Silva, T. G. F.; Santos, J. E. O.; Primo, J. T. A.; Moura, M. S. B. de; Silva, S. M. S. e; Silva, M. da C.; Souza, L. S. B. de. Soil water dynamics in an area cultivated with forage palm clones. In: Brazilian Congress of Meteorology, Gramado. Proceedings... Gramado: SBMET, 2012.

Silva, W. A.; Santana, J. S.; Silva, C. M; Nunes, A. A. Crop coefficient regionalisation applied to irrigated agriculture planning in the State of Maranhão-Brazil. Engenharia Agrícola, v.37, n.5, 2017b.

Sousa, G. G. de; Marinho, A. B.; Albuquerque, A. H. P.; Viana, T. V. A.; Azevedo, B. M. Initial growth of maize under different concentrations of bovine biofertiliser irrigated with saline water. Revista Ciência Agronômica, v.43, p.237-245, 2012.

Sousa, J. R. M.; Gheyi, H. R.; Brito, M. E. B.; Lima, G. S de. Cell membrane damage and chlorophyll pigments in citrus under saline water and nitrogen fertilisation. Irriga, v. 22, p. 353- 368,

2 017.

Souza, A. E.; Nascimento, L. C.; Souza, B. O. de. Principal components of the intensity of squamous rot on prickly pear plantations in the semiarid region of the state of Paraíba, Brazil. Revista Caatinga, v.30, n.2, p.370-376, 2017.

Souza, M. V. P.; Sousa, G. G.; Silva Sales, J. R.; Costa Freire, M. H.; Silva, G. L.; Araújo V. T. V. Saline water and cattle and goat manure biofertilisers on soil salinity, growth and physiology of broad bean. Brazilian Journal of Agricultural Sciences, v.14, n.3, p.5672, 2019.

TAIZ, L. et al. Plant physiology and development. 6.ed. Porto Alegre: Artmed, 2017.

Ting, I. P. Environmental biology of agaves and cacti. Science, v.242, p.948-950, 1988.

Toledo, J. V.; Changes in evapotranspiration of sugarcane cultivars under different levels of salinity applied in fertigation. Revista Engenharia Agrícola, v.37, n.5, 2017.

Vasconcelos, A. G. V.; Lira, M. A.; Cavalcanti, V. L. B.; Santos, M. V. F. dos.; Willadino, L. Selection of forage palm clones resistant to the carmine mealybug (Dactylopius sp). Revista Brasileira de Zootecnia, v.38, p.827-831, 2009.

Youngs, E. G. Hydraulic conductivity of saturated soils. In: SMITH, K.A.; MULLINS, C.E. (Ed). Soil analysis: physical methods. New York: Marcel Dekker, 1991. Chap.4, p.161-207.

MIX
Papier aus verantwortungsvollen Quellen
Paper from responsible sources
FSC® C105338

Printed by Books on Demand GmbH, Norderstedt / Germany